BEGINNER'S GUIDE TO Minerals & ROCKS

BEGINNER'S GUIDE TO Minerals & ROCKS

Dr. Joel Grice
Photographs by Ole Johnsen

Beginner's Guide to Rocks and Minerals
Text and photographs by R.A. Gault copyright © 2010 Canadian Museum of Nature
Photographs by Ole Johnsen copyright © 2010 Ole Johnsen

All rights reserved. No part of this book may be reproduced in any manner without the express written consent of the publisher, except in the case of brief excerpts in critical reviews and articles.
All inquiries should be addressed to:

Fitzhenry & Whiteside Ltd.
195 Allstate Parkway,
Markham, ON L3R 4T8

In the United States:
311 Washington Street,
Brighton, Massachusetts 02135

www.fitzhenry.ca godwit@fitzhenry.ca

Fitzhenry & Whiteside acknowledges with thanks the Canada Council for the Arts, and the Ontario Arts Council for their support of our publishing program. We acknowledge the financial support of the Government of Canada through the Book Publishing Industry Development Program (BPIDP) for our publishing activities.

Library and Archives Canada Cataloguing in Publication
Grice, Joel
Beginner's guide to rocks and minerals / Joel Grice;
photographs by Ole Johnsen
Includes bibliographical references and index.
ISBN 978-1-55041-584-1 (bound).—ISBN 978-1-55455-096-8 (pbk.)
1. Rocks. 2. Minerals. I. Title.
QE432.G75 2009 552 C2008-907428-9

United States Cataloguing-in-Publication Data
Grice, Joel.
Beginner's guide to minerals & rocks / Joel Grice;
photographs by Ole Johnsen
[216] p. : col. photos. ; cm.
ISBN: 978-1-55041-584-1
ISBN: 978-1-55455-096-8 (pbk.)
1. Rocks—Handbooks, manuals, etc. 2. Minerals—Handbooks, manuals, etc. I. Title.
552 dc22 QE366.8G753. 2009

Cover and interior design by Kerry Designs
Printed and bound in Hong Kong, China

1 3 5 7 9 10 8 6 4 2

To Dahlia Tanasoiu for her contribution in the early planning and outline of this book along with Wendy McPeake. My friends and colleagues: Dorrit Johnsen edited all of the digital photographs; Bob Gault took a number of the mineral and rock photographs; Donna Naughton took a number of field photographs and prepared the Geological Regions map along with Alan McDonald; Nancy Boase and Michel Picard organized the specimens for photography and prepared all of the locality information. Lorraine Brown helped by editing the first draft of the mineral and rock descriptions.

Contents

Introduction

How to Use This Book

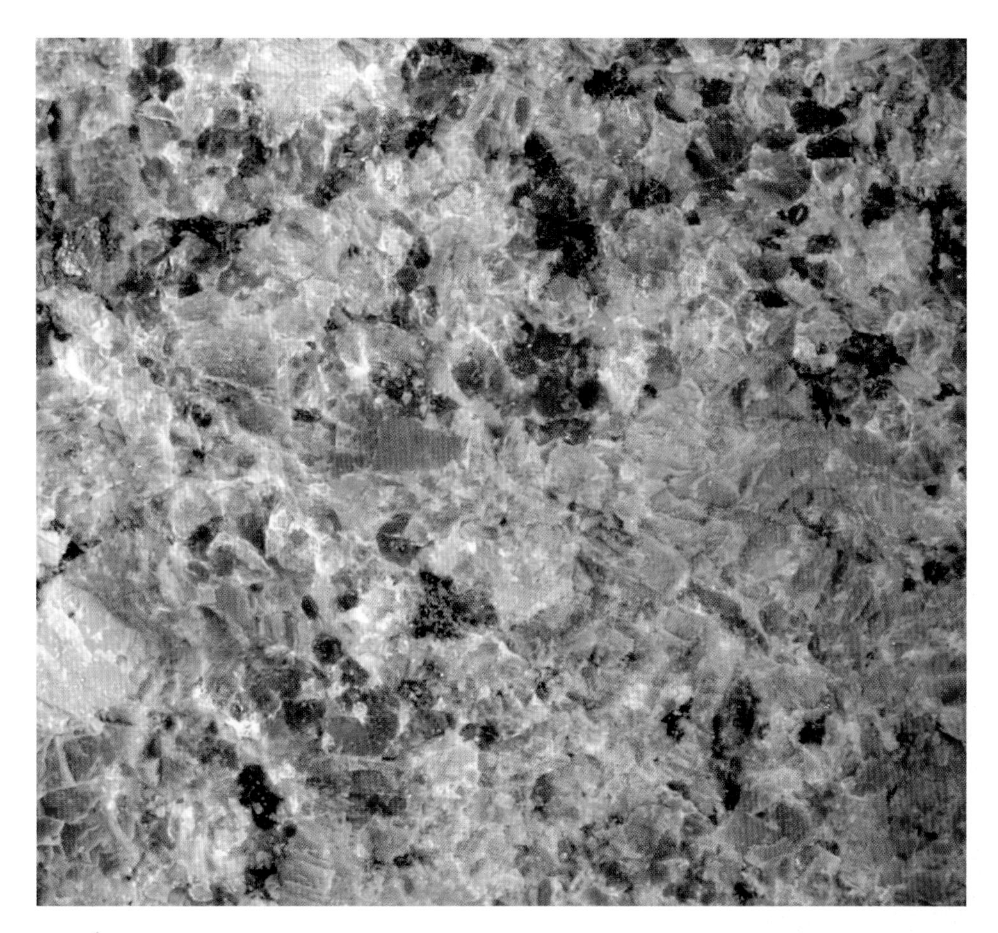

Rocks

Introduction

Mineral and Rock Collecting

For most of us collecting is a natural instinct; humans are gatherers. There is something truly satisfying about collecting natural objects like insects (zoology), leaves (botany), rocks (petrology), or minerals (mineralogy). All of us are certainly attracted to plants and animals but collecting them poses problems with killing something and then trying to preserve it. Minerals and rocks do not have these inherent problems. Minerals have their own special allure, with their beauty and their variety of shapes, forms, and colours. Finding them is very much like a treasure hunt that none of us can resist. Collecting takes you out into nature where you can observe biology without interfering with it. Mineral collecting develops your appreciation of our world and how it can work for you. An added bonus, not to be ignored, is that mineral collecting gives you a healthy, active, outdoor life.

Mineral or Rock?

Mineralogists would define a mineral as one or more chemical elements, with a crystal structure and formed through a geological process.

Every time we attempt to define something in nature there will be exceptions or complications. But let us not debate these. It is worth noting, however, the non-scientific, and somewhat incorrect, use of the term "mineral" in "mineral water" or "vitamins and minerals." There is also widespread misuse of mineral names such as diamond, quartz, and ruby for substances synthesized in a laboratory. These should be clearly described as synthetic diamond, synthetic quartz, or synthetic ruby so there is no misunderstanding.

A rock is usually an aggregate of two or more minerals that are cemented or fused together. A few rocks such as limestone or quartzite may be composed of one mineral species but they are still an aggregate of many grains or crystals cemented together.

Origins of Minerals and Rocks

Minerals grow as crystals from material presented to the mineral surface. This material must contain the right chemical elements, free to arrange themselves correctly in the atomic structure.

Rocks are an essential component of our solar system. They are grouped according to how they were formed.

Igneous rocks are derived deep in the earth where temperatures are high enough to form a liquid or magma. As the magma rises to the surface it cools and separates into different fractions. The more mafic fractions with darker iron- and magnesium-rich minerals crystallize first and the felsic fractions with lighter coloured minerals richer in silicon and aluminum crystallize closer to the surface at cooler temperatures. Igneous rocks may be intrusive, crystallizing at depth, or extrusive if they come to the surface of the earth and

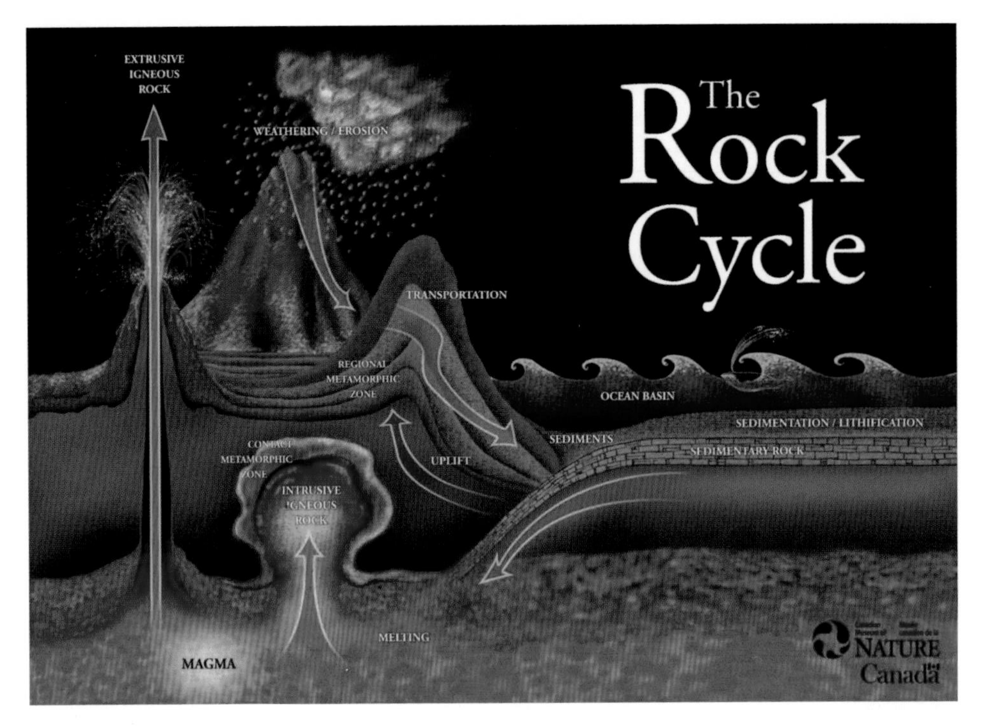

erupt as a volcano. An example of intrusive rock is granite while an extrusive example could be basalt.

Sedimentary rocks are derived from the abrasion or erosion of other rocks forming loose particles that collect and are later compacted and cemented together (lithification). Examples are sandstone and limestone.

Metamorphic rocks are derived by subjecting igneous or sedimentary rocks to increases in heat and pressure or by the action of permeating fluids that alter the chemical composition. Contact metamorphism is changes brought about adjacent to a heat source like a magma chamber (for example, limestone heated and recrystallized to marble). Regional metamorphism is large-scale alterations of rock due to both temperatures and pressure, for example the building of a mountain chain that produces mica schist from previous sedimentary rocks.

The formation of rocks can be looked at as a continuous cycle.

Field Equipment

Keep your field equipment simple but of high quality. Below I have laid out what I carry into the field for mineral collecting. The tools will be found in most good hardware stores.

Safety glasses or goggles are your single most important piece of equipment. Hitting rocks or minerals sends piercing shards in all directions. If you are collecting adjacent to another person this danger is compounded as fragments from their work can also find your eye.

Hammer: I prefer a one-handed sledge of one-half or one kilogram weight. The geology pick

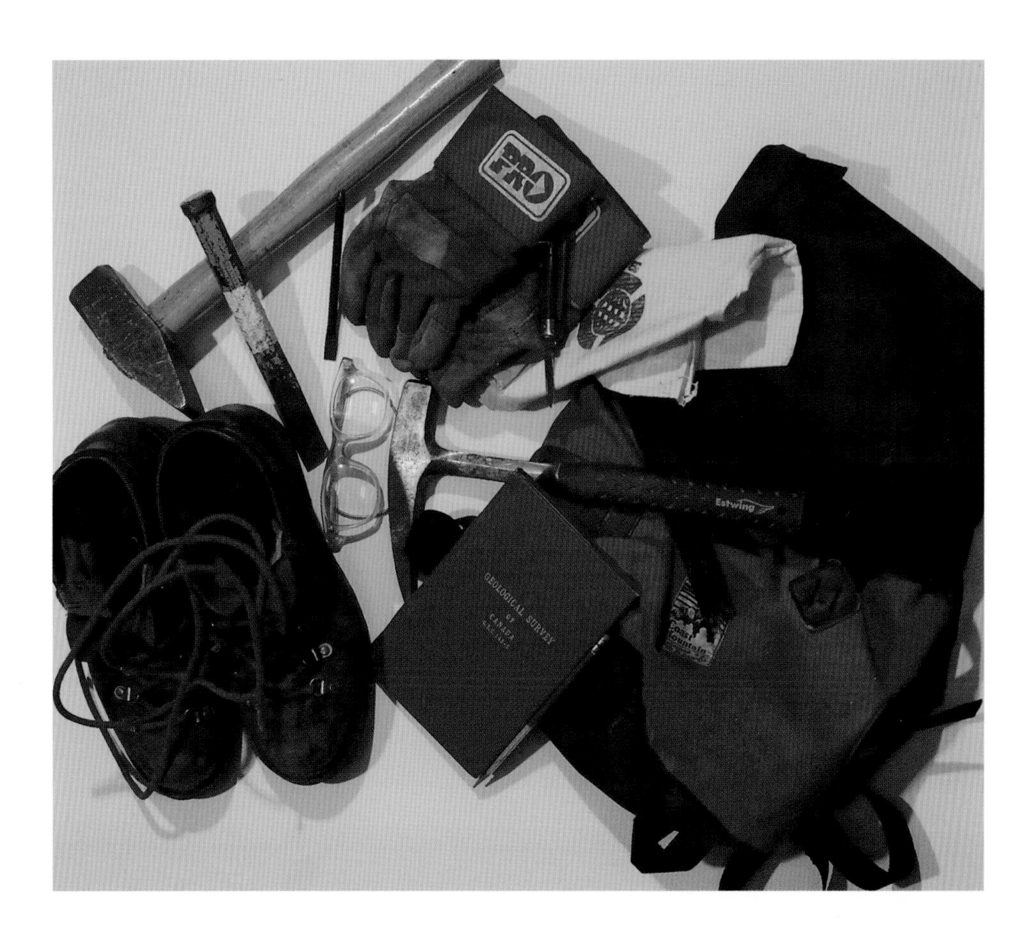

hammer looks professional and is certainly fine for breaking off a chunk of rock, but if you are collecting more delicate mineral specimens a hammer plus a chisel gives you much more control.

Chisel: Buy two if you can afford it, a small (1 cm along the splitting edge) and a large (2 or 3 cm on edge). Buy good quality hardened steel as they wear quickly. Rocks are hard.

Magnifying lens: These can be expensive but initially you don't need the best, particularly if you have good eyesight.

Notebook and pencil to keep track of where you collected and what you collected if you can identify it in the field. The locality information is the most important note. After all, you can always identify the mineral later but if you don't know where you got it, you have lost the most important information.

Old newspaper for wrapping specimens.

Backpack or carrying bag as you want to keep your hands free.

Gloves and boots: Good heavy ones are often nice to have to save some cuts and bruises from flying rock fragments. I know geologists who have suffered broken toes and injured feet because of falling rocks and improper footwear.

Compass and maps, or if you are being more adventurous, GPS.

A pocket knife can be useful for determining hardness or making your lunch.

Where to Collect

It's a big world; where do we start our collection?

Often this just happens before you know what you're getting into. Rocks, and the minerals in them, are exposed everywhere, along road cuts, river banks, cliffs, ridges, quarries, and excavations. Localities can be found in books or on the Internet.

Watch out for falling rocks.

Obey property ownership rights.

Home Equipment

At home you need little else for equipment. Your collection should grow with thought. It is easy and totally useless to just keep gathering. It takes room, becomes messy and serves no purpose. The image below shows what I had at home for identifying my collection. Do not carry acid, sharp objects, or expensive tools in the field. You will lose or damage them. It is not always possible, or necessary, to identify everything in the field.

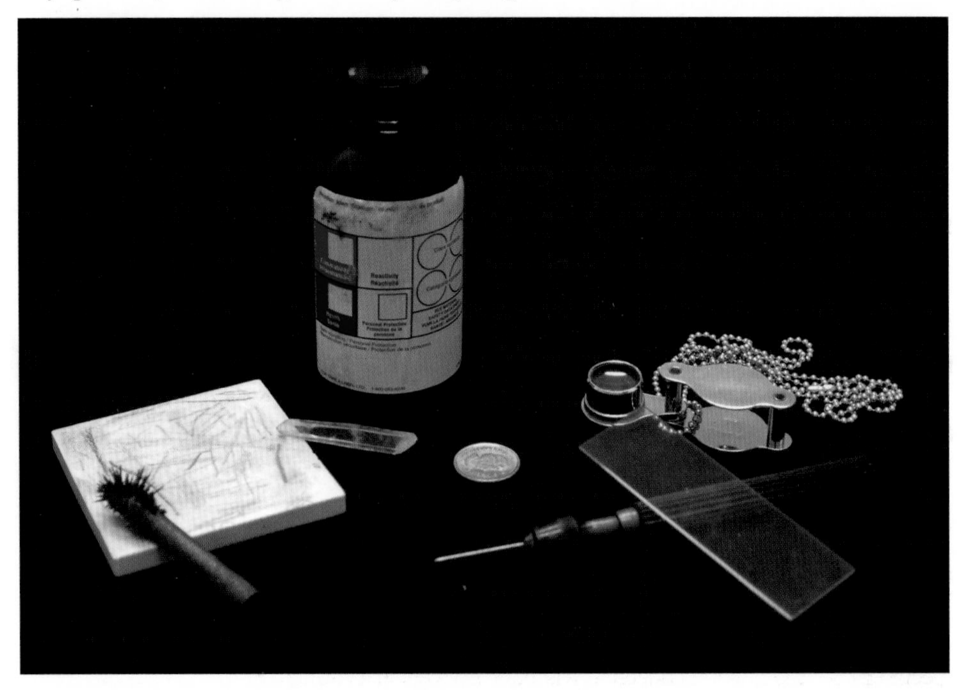

Acid: Dilute hydrochloric (HCl) with 4 parts water mixed in 1 part HCl. If this acid isn't available use vinegar. One or two drops of this solution on a mineral or rock specimen will fizz (carbon dioxide gas) if it contains carbonate in its chemical composition in minerals such as calcite, dolomite, and aragonite, or rocks such as limestone, dolostone, and marble.

Streak plate: Use a piece of unglazed porcelain, such as the back of a piece of bathroom or kitchen tile. Remember with this test you are merely looking at the colour of the powder so grinding a little piece of mineral is sufficient.

Hardness (Mohs hardness scale): Testing tools include fingernail (hardness 2½), penny (hardness 3), a knife or needle (hardness 5½), glass (hardness 6), and a piece of quartz (hardness 7).
Hand lens: A 10-power (10x) lens is sufficient.

Cataloguing Your Collection

Early on in your collecting it is wise to have some idea as to what and why you are collecting. You may wish to collect for beauty, in which case you need some way of displaying your samples. Many of us collect for study which requires a little more organization. Critical to this organization is a catalogue and storage.

The single most important item in a museum is the catalogue. It should be given the same care and respect in a private collection. In a card file, book, or computer you should record the information listed below. Without a catalogue the collection can lose much of its value.

The catalogue registers mportant pieces of information as well as any notes you may wish to add:

Catalogue number: Use a consecutive numbering system based on order of acquisition. This number is put on the specimen.

Mineral or rock name: Do the best you can to identify your specimen using the information in the section "How to Use This Book". Remember if you make a mistake in identification it really is not that serious; it can be changed at any time and your specimen can be reclassified within the collection.

Locality: This is the single most important piece of information. In most cases if this is not registered it cannot be determined later. Be as precise as possible. If you collected the specimen, record the location with enough information that you or anyone else could find the place again.

How obtained or provenance: Was it collected, exchanged, a gift, purchase, (and price)? Record as much information as you can.

Other: Less important information can go here, such as other minerals you found associated with the specimen; geological setting; price if you bought it; how many pieces in the event that you have more than one with the same number.

Storing Your Collection

A collection is only as good as its organization. You may display some of your more attractive specimens but most of the collection will be in drawers in a cabinet.

- Each specimen receives a number corresponding to the catalogue number.
- Each sample is put in a tray that is slightly larger than the specimen. Line the tray if necessary for protection.
- Place a label in each tray with catalogue number, name, locality.
- Place the tray in appropriate drawer in cabinet.

How to Use This Book

Mineral Classification

Until one understands a great deal about chemistry and crystallography the standard textbook classification scheme is of little use. Modern texts generally follow the same system of chemical classes with subdivisions based on crystal structure. In this book I have arranged the minerals in an order that does not make the crystal-chemical classification as evident but it is the underlying principle.

It is a fundamental property of elements to have a tendency toward one type of bonding: metallic or non-metallic and, depending on the type of bonding, certain physical properties are inherent. Within this book I have split minerals into these two broad categories – metallic and non-metallic – and subdivided each category into chemical groups that have similar properties.

Metallic Minerals (25 Species)

Native elements: Minerals composed of one element. Gold, copper, silver, platinum, bismuth and graphite fall in this subdivision.

Sulphides and arsenides: Minerals with sulphur or arsenic and a metal ion. Chalcocite, bornite, chalcopyrite, covellite, pyrite, marcasite, pyrrhotite, pentlandite, stibnite, cobaltite, arsenopyrite, molybdenite, galena are included here.

Oxides: Minerals with oxygen and a metal ion, often with a semi-metallic lustre. Goethite, hematite, magnetite, pyrolusite, ilmenite, chromite are in this subdivision.

Non-metallic Minerals (51 Species)

Native elements: Minerals composed of one or two elements and with no metallic properties. Diamond and sulphur are examples found in this book.

Sulphides: Minerals with sulphur and a metal or semi-metal ion that do not have metallic properties but would best be termed sub-metallic. Sphalerite and cinnabar are examples.

Halides: Minerals with halogen elements (commonly fluorine and chlorine). Examples are halite, sylvite and fluorite, quite common minerals.

Oxide groups: Minerals that have oxygen tightly bonded in a group around a central non-metallic ion. Oxides or carbonate, borate, and phosphate groups are included. Corundum, spinel, cuprite, rutile, cassiterite, calcite, aragonite, dolomite, siderite, rhodochrosite, malachite, azurite, gypsum, anhydrite, baryte, apatite, turquoise, borax, ulexite are species in this category.

Silicates: Minerals have oxygen tightly bonded in a group around a central silicon atom. Olivine, garnet, epidote, vesuvianite, kyanite, topaz, titanite, zircon, amphibole, pyroxene, rhodonite, beryl, tourmaline, mica, chlorite, talc, clay, chrysotile, chrysocolla, quartz, feldspar, nepheline, zeolite, sodalite, lazurite are the examples included.

Characteristics or Properties of Minerals

Appearance

Colour: Beware

Colour in minerals can be very misleading. The colour of metallic minerals is more reliable than that of non-metallic minerals. For a metallic mineral the colour of the fresh surface will always be about the same. Hence, one has only to be aware of whether the surface is fresh or tarnished (the way a penny can be shinny or dark). Tarnish can help with identification but, if it is mistaken for the actual colour of the mineral, errors in identification can be made. For non-metallic, transparent, or translucent minerals, colour may be deceiving, since small changes in chemistry or the origin of the mineral may change the colour drastically, for example, tourmaline, garnet, topaz, quartz.

Streak: Colour of Powder

Streak is the colour of the mineral powder and it is far more consistent than the colour of the entire specimen. Most commonly the streak is observed by scraping the mineral across a piece of unglazed porcelain, a "streak plate." The changes in colour of streak may be subtle and one should learn these subtleties through practice on known mineral samples. For example, steel-grey hematite has a red-brown streak, while steel-grey molybdenite has a grey streak.

Lustre: Surface Reflection

Lustre describes the surface appearance of a mineral – how it reflects light. Mineralogists have adopted many terms in an effort to describe this feature as it is important in identification. In this book, the initial step in identification is deciding whether the mineral is metallic or non-metallic.

A metallic mineral is opaque and shiny like metal, e.g. copper, silver, gold.

A non-metallic mineral is transparent or translucent with a glossy, dull or greasy appearance, e.g. quartz, siderite, diamond, gypsum.

The following is a list of other terms that describe non-metallic types of lustre. Most are self-evident but the definitions below may be helpful. Take note of the examples. Often the prefix "sub" is used when the description is not clear, e.g., submetallic, for graphite.

Adamantine: brilliant, hard, durable, e.g., diamond
Dull or earthy: not shiny or brilliant, e.g., red hematite, clay
Greasy or oily: not shiny, not bright, e.g., sphalerite
Pearly or opalescent: a depth to the light reflected, e.g., aragonite, calcite
Resinous: like gum resin secreted by trees, e.g., sphalerite
Silky: like silk in smoothness and fineness, e.g., gypsum
Vitreous: glassy, e.g., quartz, feldspar

Transparency: Passing of Light

Transparency is how light behaves when shone through the specimen. If it is transparent, light travels through the specimen with little obstruction, giving a clear view of an object behind it. If it is opaque, no light passes through the specimen no matter how thin it is. If it is translucent, light passes through the specimen but the object behind it cannot be seen clearly.

Remember that a transparent mineral can be colourless or coloured. The term "clear" does not imply that the mineral is without colour.

Habit

"Tout est trouvé" *(All has been discovered.)* Abbé René Just Haüy made this proclamation after many years of studying crystals and their growth features. His insight into the laws of nature showed how crystals consist of identical building blocks or cells stacked in three dimensions. It would be another two hundred years before X-ray crystallography proved him right.

Habit is a very diagnostic feature of minerals. It combines two fundamental concepts: habit – the appearance resulting from crystal growth, and form – the shape of a single crystal.

Habit describes the characteristic appearance of the crystal or an aggregate of crystals, i.e, how they grew. Terms used by mineralogists to describe habits of crystals are:

Acicular: needle-shaped

Bladed: long, thin plate like a knife blade

Botryoidal: rounded, globular (like a bunch of grapes)

Dendritic: tree-like or fern-like

Massive: compact, no crystal form

Parallel growth: crystals that grew parallel to each other but are not related by symmetry (see Twinned)

Prismatic: three or more faces whose intersections are parallel, like a prism

Twinned: crystals grown together and related to each other by symmetry, for example, a mirror plane.

Crystal form or the shape (morphology) of a single crystal is the most crucial aspect of mineralogy, yet I have chosen to mostly avoid it. The reason for this is the degree of difficulty in understanding the concepts. When a crystal is growing it obeys certain laws of physics which require the atoms to bond in a well-defined manner. This results in seven basic crystal systems, each defined by specific symmetry elements. Crystal form is important for an advanced mineral collector to identify more difficult mineral species. (See the chart on p. 9 for crystal forms.)

Physical Properties of Minerals

Hardness

One of the more useful characteristics for identifying a mineral is its hardness. How difficult it is to scratch a mineral depends on the fundamental physical property of how tightly the atoms are bonded together within the crystal structure. This property may be quantified precisely with complex experimental techniques but for our use there is a simple relative scale of hardness that was developed by Friedrich Mohs in 1822. He chose ten minerals, each progressively harder than the next, thus providing a relative scale.

The Mohs scale of hardness (with 1 being the softest) is as follows: 1 talc, 2 gypsum, 3 calcite, 4 fluorite, 5 apatite, 6 feldspar, 7 quartz, 8 topaz, 9 corundum, 10 diamond.

For field identification it is sufficient to broadly differentiate hardness into three groups based on softer than a finger nail (Mohs 2), harder than a finger nail but softer than a knife blade (Mohs 5), and lastly harder than a knife blade.

Crystal Forms

Crystal form is: cube {a}.

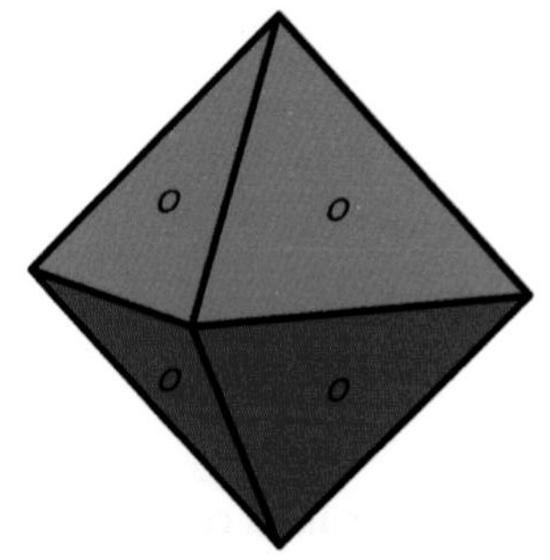

Crystal form is: octahedron {o}.

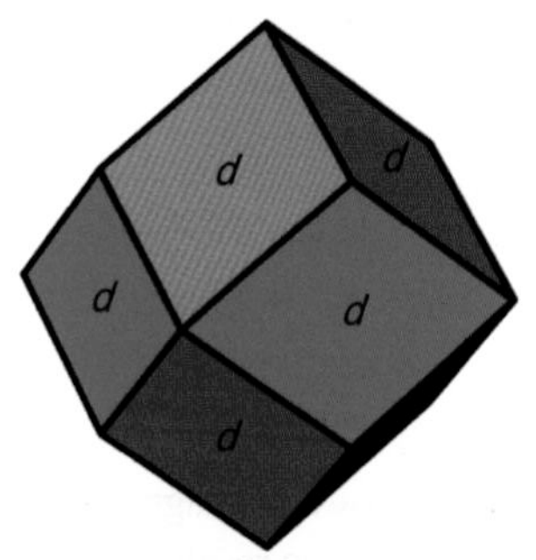

Crystal form is: rhombic dodecahedron {d}.

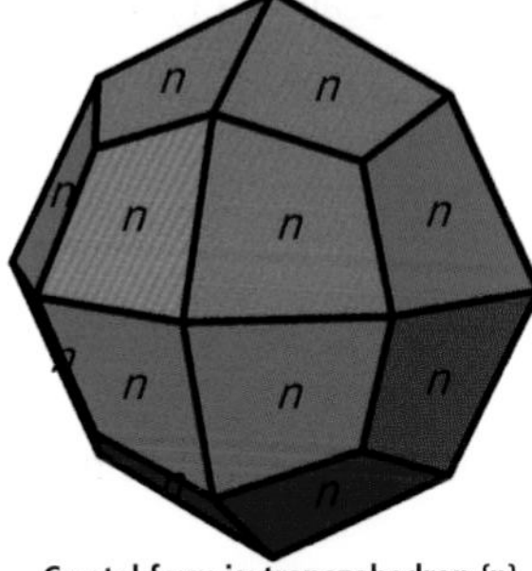

Crystal form is: trapezohedron {n}.

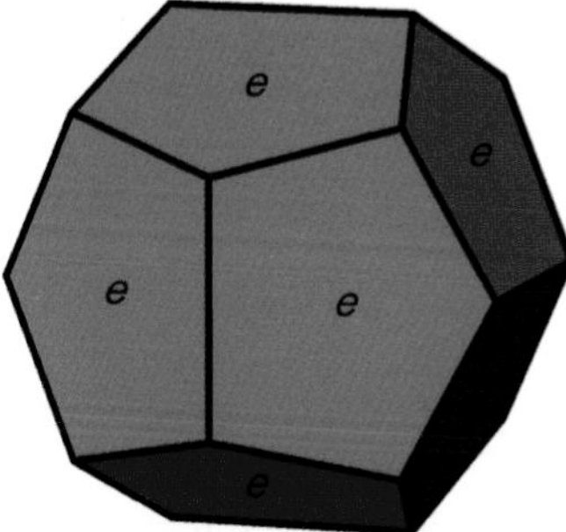

Crystal form is: pyritohedron or pentagonal dodecahedron {e}.

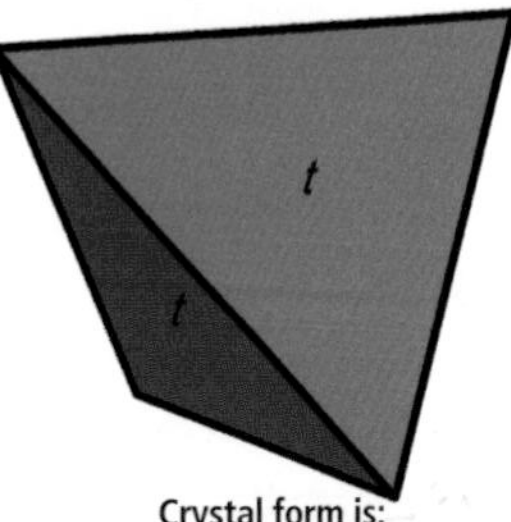

Crystal form is: tetrahedron {t}.

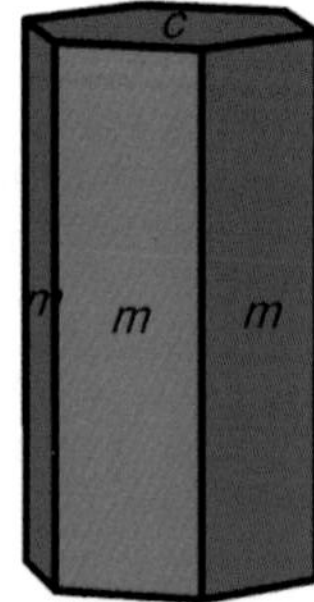

Crystal forms are: hexagonal prism {m} and pinacoid {c}.

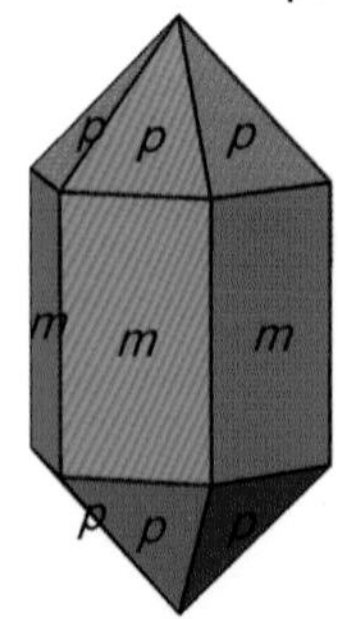

Crystal forms are: hexagonal prism {m}, hexagonal dipyramid {p}.

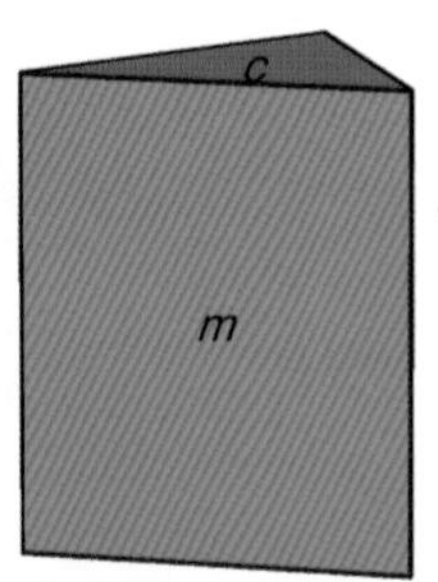

Crystal forms are: trigonal prism {m} and pinacoid {c}.

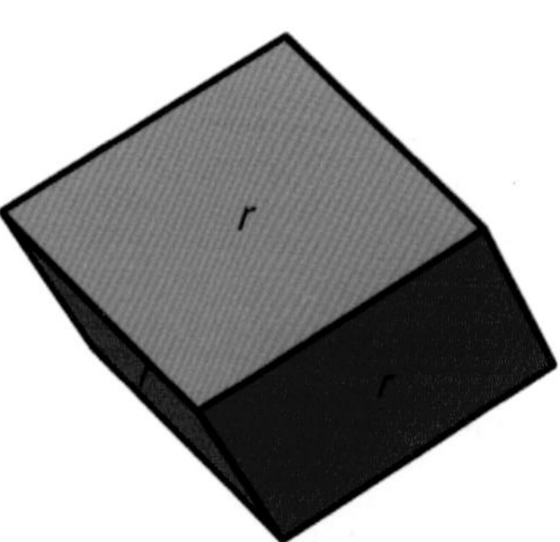

Crystal form is: rhombohedron {r}.

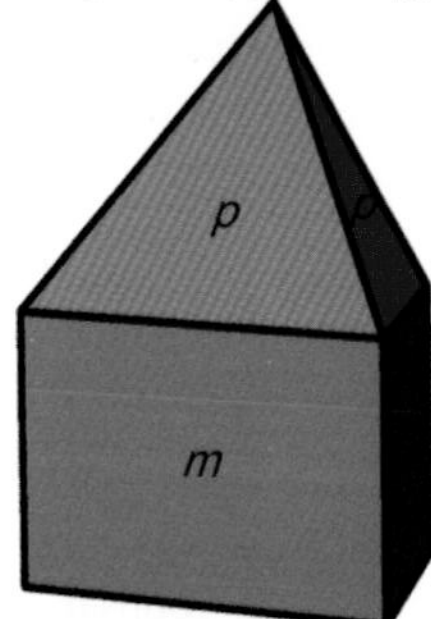

Crystal forms are: trigonal prism {m}, trigonal pyramid {p} on a basal pinacoid {c}.

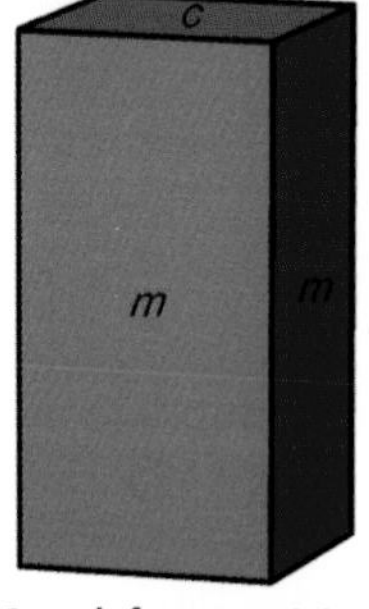

Crystals forms are: tetragonal prism {m} and pinacoid {c}.

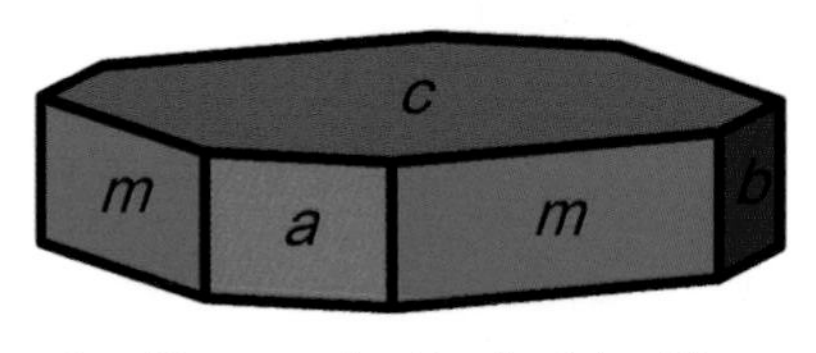

Crystal forms are: rhombic prism {m} and three pinacoids {a}, {b}, and {c}.

If possible do the hardness test on a portion of the specimen that is not obvious, as it spoils the appearance of your specimen.

It may be of interest that on an absolute scale of hardness corundum is actually two thousand times harder to scratch than talc, whereas on the Mohs relative scale it appears to be only nine times harder.

Density: Measuring Heft

Before lifting an object, a person has a feeling as to what that object will weigh. This "heft" is an approximate measure of density: the weight of a specific volume of material. With practice one should be able to differentiate between minerals more or less denser than common quartz or feldspar (2.6 g/cm^3). Baryte and fluorite are much heavier with densities of 4.5 g/cm^3 and 3.2 g/cm^3 respectively. Gold, the most dense mineral, measures 19 g/cm^3, and ice, the least dense, has a density of 0.9 g/cm^3. It may be helpful to remember the density of water at 1.0 g/cm^3, thus ice floats.

Breakage: Something to Avoid

It is not acceptable to break your mineral sample intentionally just to see how it breaks. If you carefully inspect the sample, often it will give you clues as to how it breaks. This can be helpful in identification. You want to know how easily it breaks and the shape or form of the break.

Within a crystal, atoms bond together by electrical force of attraction. Along planes of weak bonding the mineral tends to break or cleave. This planar separation is called **cleavage**. It may happen on one or several planes. For example there is one plane in mica, chlorite, topaz; there are two planes in amphibole and pyroxene and three planes in fluorite and calcite.

If the mineral does not have cleavage it has a rough surface of breakage called **fracture**. Fractures may be described as conchoidal (rounded), hackly (jagged), or splintery (shards).

How to Identify a Mineral

Mineral identification is not easy. It is a skill and like all skills it requires practice. Advanced mineral collectors can identify specimens in an instant without even touching them. In many cases they can even tell you where the specimen came from. This degree of skill takes years of practice to develop. Even for a professional mineralogist there are many cases where only sophisticated equipment like X-ray diffraction and chemical analyses will definitively determine which of the 4,000 possible mineral species you have.

This book contains the most common minerals. The method below will help you identify your specimens.

Identification Method

First determine whether your specimen is metallic or non-metallic (see p. 7). This observation will lead you to the Metallic Mineral Key (p. 11) or the Non-Metallic Mineral Key (p. 12).

Next, use your streak plate (see p. 7) to determine the streak colour. Find the colour on the Metallic or Non-Metallic Key.

The next step is to determine the hardness. If you can scratch your specimen with your fingernail, it is soft (Mohs 1 – 2). If you can scratch it with a knife blade, your specimen is considered to be moderately hard (Mohs 2½ – 5½). If you cannot scratch your mineral with a knife blade, it is hard (Mohs greater than 5).

Using the key, you will find several choices for your specimen. The page numbers refer to the descriptions and pictures in the book. Check each page number until you find the one that most closely matches your specimen.

Colour is one of the most deceiving properties of minerals. Many species have several quite different colours so make sure you check each entry.

Metallic Mineral Key

Streak Colour	Degree of Hardness		
	Soft (Mohs 1 – 2½)	**Moderate** (Mohs 2½ – 5½)	**Hard** (Mohs more than 5½)
Black	covellite (dark blue, black), p. 36 pyrolusite (steel grey with bluish tint), p. 82 graphite (steel grey, black), p. 30 stibnite (lead grey), p. 62	chalcopyrite (brassy yellow), p. 42 bornite (bronze), p. 38 pyrrhotite (bronze yellow), p. 50 chalcocite (dark grey), p. 34	magnetite (black), p. 74 pyrite (pale brass yellow), p. 46 arsenopyrite (silver-grey), p. 54
Grey	galena (silver grey, dull lead), p. 56 molybdenite (lead grey with bluish tint), p. 60 stibnite (lead grey), p. 62	platinum (silver-white), p. 26 chalcocite (dark grey), p. 34	cobaltite (silver-grey), p. 64 marcasite (pale brass yellow), p. 48
Silvery White	bismuth (reddish silver-white), p. 28	silver (silver-white), p. 22	
White		silver (silver-white), p. 22 platinum (silver-white), p. 26	titanite (brown to black), p. 176
Red-Brown		copper (reddish to brown), p. 20 cuprite (dark red, black), p. 116 chromite (black), p. 80	hematite (steel grey to red), p. 66 ilmenite (brownish black, black), p. 66 rutile (brown, dark red brown), p. 118 chromite, p. 80
Yellow		gold (gold yellow), p. 18 pentlandite (light bronze yellow), p. 52 goethite (yellow, red), p. 72	

Non-metallic Mineral Key

Streak Colour	Degree of Hardness		
	Soft (Mohs 1 – 2½)	**Moderate** (Mohs 2½ – 5½)	**Hard** (Mohs more than 5½)
Black	graphite (black), p. 30 pyrolusite (grey to black), p. 82		
Greyish white			amphibole (green to almost black), p. 196 cassiterite (dark reddish brown, black), p. 120 epidote (green), p. 180
White	borax (colourless, white), p. 160 chlorite (pale green), p. 208 clay (white, brownish), p. 212 gypsum (white), p. 124 halite (white), p. 100 sulphur (yellow), p. 90 sylvite (colourless, white, reddish), p. 104 talc (white, pale green), p. 210 ulexite (colourless, white), p. 162	anhydrite (white or grey), p. 122 aragonite (white, variable), p. 146 calcite (white, variable), p. 138 dolomite (grey, pink), p. 280 baryte (white, pale blue), p. 132 chrysotile (green, white, yellow), p. 214 apatite (green, reddish brown), p. 128 fluorite (colourless, blue, green), p. 106 kyanite (blue), p. 170 mica (white, brown, black), p. 204 rhodochrosite (red, pink), p. 152 rhodonite (red, brown-red), p. 202 siderite (brown), p. 150 titanite (brown black), p. 176 zeolite (colourless, white, yellow), p. 238	beryl (white, variable), p. 184 cassiterite (dark reddish brown, black), p. 120 corundum (grey, blue), p. 112 diamond (colourless), p. 86 feldspar (white, variable), p. 226 garnet (red, orange, green), p. 166 nepheline (white, grey), p. 232 olivine (green, yellowish), p. 164 pyroxene (green to black), p. 192 quartz (colourless, white, variable), p. 218 spinel (red, brown, black), p. 110 titanite (brown black), p. 176 topaz (blue, whiteyellow), p. 172 tourmaline (black, brown, variable), p. 188 vesuvianite (green, brown), p. 182
Blue		azurite (dark blue), p. 156 chrysocolla (blue green), p. 216	lazurite (blue), p. 236 sodalite (blue), p. 234
Green	chlorite (pale green), p. 208	malachite (green), p. 158	turquoise (greenish blue), p. 136
Red, Red-brown	cinnabar (red), p. 96	cuprite (red to almost black), p. 116 sphalerite (brown, yellow), p. 92 goethite (yellow, red), p. 72	hematite (red, black), p. 68 ilmenite (red to brownish black), p. 78 rutile (reddish brown), p. 118

How to Identify a Rock

Only a few important rocks are covered in this book and they are all in one identification key (below). They are the most common and they reflect very differing origins. A better determination of special rock types requires a petrographic microscope.

Identification Method

First, determine which of the three rock types you have:

Igneous rocks have no fossils or bedding layers (sedimentary) or folds (metamorphic). Lavas may be filled with holes or glassy patches. Interlocking crystals that are randomly distributed.

Sedimentary rocks have individual grains or fragments that are poorly held together. Fossils may be present. Quartz and calcite are common minerals.

Metamorphic rocks characteristically have visible bandings of minerals, often wavy or foliated. Individual minerals are welded together.

Second, determine if your rock is coarse-grained or fine-grained.

Third, determine the relative degree of colour of the rock. Is it light, medium or dark in colour?

Check the information in the Rock Key to the right, and using the page numbers find your specimen.

Rock Key

Igneous Rocks

Colour	Grain size	
	Coarse	Fine
Light	granite (p. 248) syenite (p. 254)	rhyolite (p. 256)
Medium	diorite (p. 250)	andesite (p. 260)
Dark	gabbro (p. 252)	basalt (p. 262) obsidian (p. 266)

Sedimentary Rocks

Colour	Grain size	
	Coarse	Fine
Light	conglomerate (p. 268) breccia (p. 270) sandstone (p. 272)	limestone (p. 276) siltstone (p. 274)
Medium	conglomerate (p. 268) sandstone (p. 272)	dolomite (p. 142)
Dark	breccia (p. 270) conglomerate (p. 268)	shale (p. 282)

Metamorphic Rocks

Colour	Grain size	
	Coarse	Fine
Light	marble (p. 294) granite gneiss (p. 290) quartzite (p. 296)	quartzite (p. 294)
Medium	quartzite (p. 296) schist (p. 286)	quartzite (p. 286)
Dark	amphibole gneiss (p. 280)	slate (p. 284) serpentinite (p. 298)

Geological Regions of Canada

Those fortunate people who have travelled part or all of the six-thousand-kilometre breadth of Canada will have noted stunning differences in the surface features of the countryside. These features make up the physiographic and hence geological regions of the country. Very broadly, there are two great parts: a central core of very ancient and erosion-resistant rock called the Precambrian Shield, and a younger skirting ring of layered rocks.

There are eight physiographic regions in Canada:

1 The Canadian Shield, a rugged country of hard, resistant rock cut by innumerable lakes and rivers, comprises the core of the Continent.
2 The St. Lawrence Lowlands form the southernmost portion of Canada.
3 Great expanses of flat prairie and sedimentary rocks mark the Interior Plains.
4 Below the Innuitian Region lie the Arctic Lowlands.
5 Inside the Shield lie the Hudson Bay Lowlands.
6 The worn mountains and hills of the Appalachians extend along the eastern margins of Canada.
7 The older mountains of the Innuitian Region make up the country's northernmost territory.
8 The rough, spectacular mountain ranges of the Cordilleran Region rim the west side of the continent.

These eight physiographic regions depend directly on the underlying geology. Each region has a distinct suite of rocks, differing from each other in age and provenance. Although the details of any region are complex, some general statements may lead to a better understanding of the specific mineral localities listed in subsequent descriptions.

Canadian Shield

The Canadian, or Precambrian, Shield constitutes the stable core of the North American continent. These ancient rocks vary in age from a billion to over four billion years, making them some of the oldest rocks in the world. They formed deep in the primeval crust of planet Earth and lie exposed today after billions of years of erosion by rain, ice, and wind. The original rocks underwent tortuous changes in temperature and pressure, producing a series of highly metamorphosed, coarsely crystalline rocks.

The Shield, named for its shape, is somewhat like a saucer, being higher on the outer edges than in the centre. It constitutes approximately half of Canada's land surface and boasts one of the most productive mining areas in the world. The Shield's rich deposits of copper, nickel, iron, lead, zinc, gold, silver, cobalt, uranium, platinum, titanium, and molybdenum support Canada's economy.

Platform Regions

The Platform, a series of flat, sedimentary rocks, overlie the Canadian Shield in a broad collar around its southern, western, and northern rims. This geological feature is evident in the **St. Lawrence Lowlands**, the **Interior Plains** of the Midwest, the **Arctic Lowlands** to the north, and the **Hudson Bay Lowlands** which infills the middle of the Shield. The sediments from the

Shield, Cordilleran, or Appalachian regions were deposited in seas that formerly covered the Shield. The sediments formed a thin veneer on the Shield, except in the Interior Plains where they are several kilometres thick. The Interior Plains produce most of Canada's petroleum and natural gas as well as potash, salt, gypsum, and limestone. During the ice age, soils ground from the Shield by the ice were deposited on the Platform. In the southern prairies and the St. Lawrence valley, this fragile layer of sediments produced fertile agricultural land. Unfortunately, much of this is now covered by urban sprawl.

Appalachian Region

In North America, the old, worn Appalachian Mountains stretch three thousand kilometres, from Newfoundland to Alabama. They form the eastern coastline from Newfoundland to New York. Further south a flat plain of sedimentary rock separates them from the Atlantic Ocean. Formerly, the Appalachian Region consisted of a submerged trough on the edge of the Shield. Over a period of hundreds of millions of years it was filled with sediments derived from the continent and ocean. Approximately four hundred million years ago these sediments uplifted to form a range of mountains, now largely eroded away.

The Appalachian Region, including the Atlantic Provinces and southeastern Quebec, produces a good proportion of the world's asbestos as well as significant amounts of copper and zinc.

Innuitian Region

In northernmost Canada lies a region of varied topography largely covered by glaciers. The region consists of a thick series of deformed sedimentary rocks and intrusions. The northern rim contains the Grantland and Axel Heiberg Mountains, which reach elevations of 1,700 metres and protrude through ice sheets as rows of nunataks. Further south and east there is an inner belt of rugged, almost parallel ridges that cross the entire region. The southernmost portion of the region is a plateau of broad, flat-topped ridges cut by deep ravines or fiords.

Cordilleran Region

Canada's newest mountains are part of the Cordilleran Mountain Region, which spans half the world – eighteen thousand kilometres from the Aleutian Islands to the tip of South America. This series of sedimentary and volcanic rocks heaved up approximately one hundred million years ago as a result of the collision of two huge continental plates. The theory of continental drift describes the flow of such crustal plates on the melted upper portions of the Earth's interior. The interaction resulting from the collision of the American plate and the Pacific Ocean plates remains in evidence today in the activity of the volcanoes and earthquakes along the Pacific Ocean coastline. In Canada, the Cordilleran Region displays itself in the spectacular Rocky Mountains and Coast Mountains of British Columbia.

Between the western and eastern mountain ranges, the central plateau affords some opportunity for farming. With persistent geological surveying the Cordilleran has yielded important deposits of lead, zinc, silver, copper, and gold.

Physiographic Regions of Canada

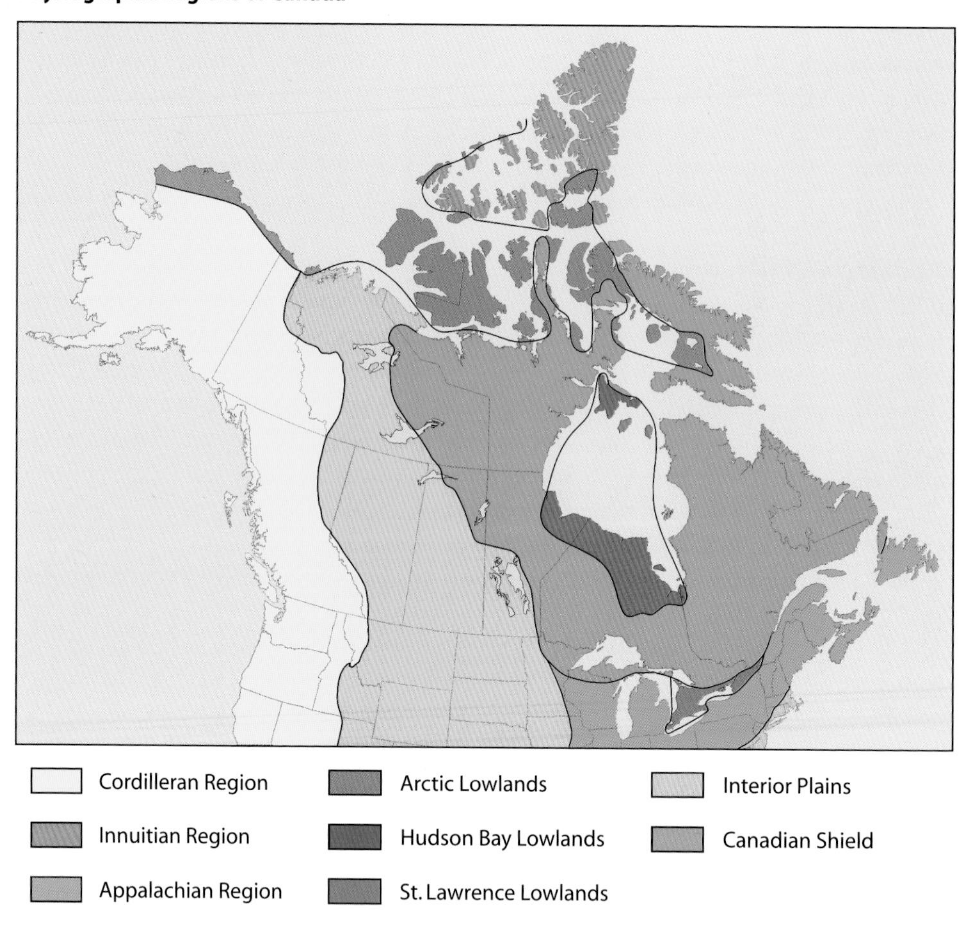

Cordilleran Region
Innuitian Region
Appalachian Region
Arctic Lowlands
Hudson Bay Lowlands
St. Lawrence Lowlands
Interior Plains
Canadian Shield

Metallic Minerals

GOLD: Au

Archaeologists believe that gold, because of its glitter, would have caught the eye of prehistoric man sooner than the other duller native metals such as copper and silver. The earliest finds of gold artefacts, dating to 4000 BCE, are from Mesopotamia, and Egypt. Although gold is neither the rarest nor the most valuable metal, it became "man's first folly" according to Pliny, 79 BCE.

Appearance

Colour: A deep yellow-gold if pure, whiter due to alloyed silver or redder due to alloyed copper.

Streak: Yellow.

Lustre and transparency: Shiny, metallic lustre and opaque.

Habit: Usually massive in lumps, or rounded grains or scales. Rare crystals are octahedral or in dendritic or fern-like aggregates.

Physical Properties

Hardness: Medium (Mohs 2½ – 3).

Density: Heavy (19.3 g/cm^3).

Breakage: Malleable and ductile, hence difficult to break, and sectile. The hackly fracture results in a jagged irregular surface like that of broken iron.

Test: Colour and the fact that it is sectile.

Gold (41795): Massive native gold in quartz. McIntyre Mine, Timmins, Tisdale Tp., Cochrane District, Ontario. Width of specimen: 10 cm

Similar Minerals

Pyrite (fool's gold) (p. 46) is pale, brassy yellow, brittle, and has a streak that is brown-black with a greenish tint.

Chalcopyrite (p. 42) is brassy yellow, brittle, with a greenish black streak.

Occurrence

It is found in placer deposits left by erosion. In consolidated rock it is found in hydrothermal deposits.

Canada's Best Localities: San Antonio Mine, Bissett, Manitoba; McIntyre Mine, Timmins, Tisdale Township, Cochrane District, Ontario; Darwin Mine, McMurray Township, Algoma District, Ontario; nuggets occur in placer deposits in the Yukon. The largest recorded from this locality is 2.64 kg.

Other Localities: The largest nugget recorded is from Carson Hill, California, USA, weighing 72.78 kg. Transylvania, Romania; Roraima, Brazil; Witwatersrand, Republic of South Africa; Ballarat, Australia.

Interesting Facts

The Anglo-Saxon word, *gold*, was first recorded in 475 CE. The chemical symbol *Au* is from the Latin *aurum*, meaning gold.

The most malleable of metals, gold can be hammered into foil a thousand times thinner than paper – so thin that light passes through it. It is very ductile; a piece of gold the size of a man's thumb can be drawn into a wire some 550 km long.

Because gold is such a good reflector of both visible and infrared light it is used as a protective coating on satellites and on astronauts' helmets to prevent blindness.

Much of the Western world adopted the "gold standard" as the basis of their currencies between the 1870s and 1914. This standard had to be abandoned in 1968 and gold was left to find its own value according to market demands. Gold prices are often quoted in dollars per ounce; this is a Troy ounce (31.10 g), which is about 10 percent heavier than the usual ounce.

The term karat is used to indicate the amount of gold present, with 24 karat being pure gold and 18 karat being 75 percent gold (18/24 x 100).

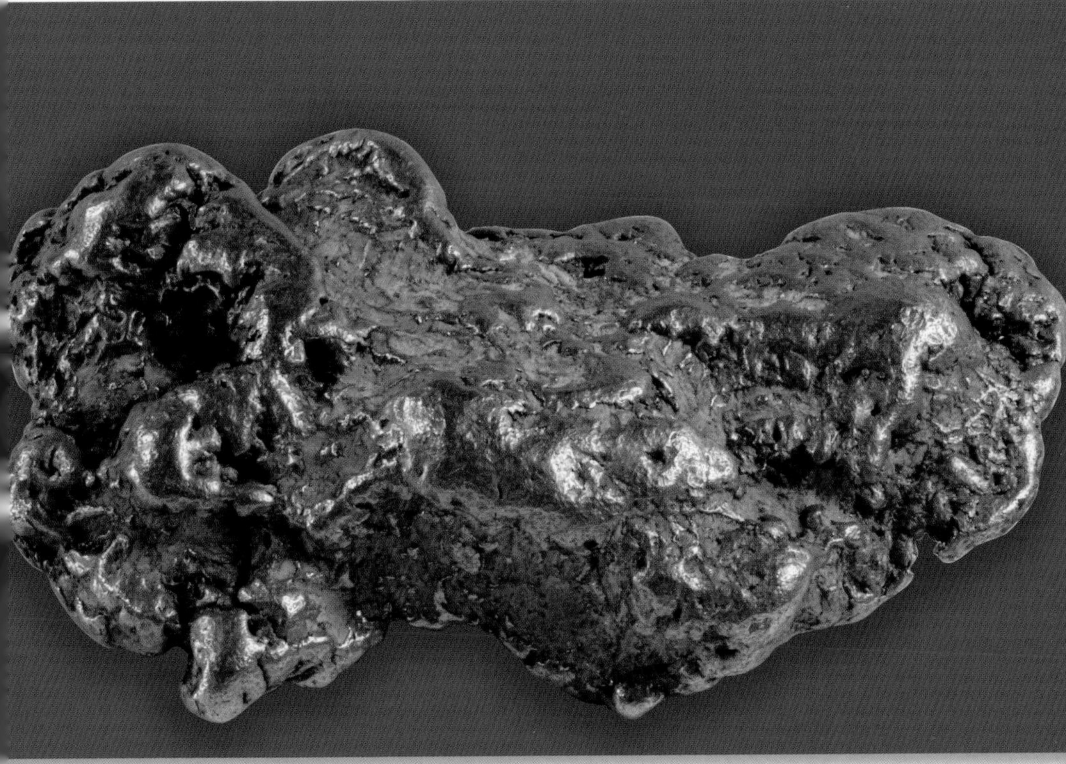

Gold (53772): This gold nugget weighing 1,450 grams is from a placer deposit in British Columbia. Width of field of view: 14 cm

COPPER: Cu

Copper is one of the few minerals that occurs in nature as an element. It is sometimes referred to as "native copper" to distinguish it from copper produced by an industrial process. Native copper was used by Neolithic man as early as 8000 BCE. There is evidence that the first Aboriginal residents of Manitoba made copper spear points as early as seven thousand years ago. Their source of copper would likely be from around the Lake Superior area.

Appearance

Colour: Copper-red to brown when tarnished.
Streak: Copper-red.
Lustre and transparency: Shiny, metallic lustre and opaque.
Habit: Usually massive. Sometimes forms twisted lumps or wires. The branching or tree-like form is termed dendritic.

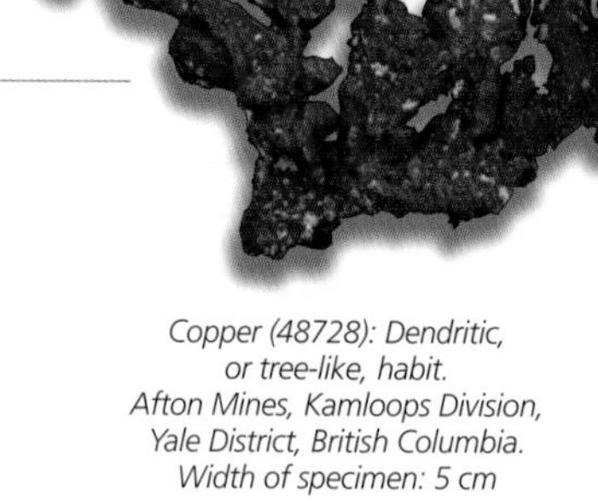

Copper (48728): Dendritic, or tree-like, habit. Afton Mines, Kamloops Division, Yale District, British Columbia. Width of specimen: 5 cm

Physical Properties

Hardness: Medium (Mohs 2½ – 3).
Density: Heavy (8.9 g/cm^3).
Breakage: Malleable and ductile, hence difficult to break, and sectile. The hackly fracture results in a jagged, irregular surface like that of broken iron.
Test: The colour of a fresh surface and malleable properties identify this mineral.

Similar Minerals

Chalcocite (p. 34) is more brittle, and has a black streak.
Cuprite (p. 116) is more brittle, deep red to black in colour, and the streak is brownish red.

Occurrence

Native copper crystallizes from hydrothermal solutions associated with basaltic rocks.

Canada's Best Localities: Craigmont Mine and Afton Mine, Kamloops District, British Columbia; Burwash Creek, Yukon Territory; Coppercorp Mine, Algoma District, Ontario; Seal Lake, Labrador, Newfoundland; Cap d'Or, Cumberland County, Nova Scotia.
Other Localities: Keweenaw Peninsula, Houton and Ontonagon Counties, Michigan, USA; Bisbee, Cochise County, Arizona, USA; Ajo, Pima County, Arizona, USA; Ray, Pinal County, Arizona, USA; Santa Rita, New Mexico, USA; Rudabanya, Hungary; Yekaterinburg Oblast', Russia; Broken Hill, New South Wales, Australia; LaPaz, Bolivia; Windhoek, Namibia.

Interesting Facts

During the Roman Empire, copper was principally mined on Cyprus, hence the origin of the name of the metal: cyprium, "metal of Cyprus," later shortened to cuprum.

In the Keweenaw Peninsula, Michigan, copper has been mined since the middle of the 19th century. There are reports of single masses of several tons. The physical properties of the mineral (malleable and sectile) hampered traditional mining techniques of drilling and blasting because it was difficult to break up areas of the mine where the mineral was concentrated. Today most copper metal is produced from mining copper sulphide ores such as those in Thompson, Manitoba, Sudbury, Ontario, and Voisey's Bay, Newfoundland.

Copper (56727): Rough crystals in breccia. Keweenaw Peninsula, Houghton Co., Michigan. Width of field of view: 7 cm

Copper's physical properties make it a good conductor of heat and electricity, rendering it useful in electronics, electrical wiring, and cookware. As it resists corrosion, it is used in roofing and statuary. Alloyed with nickel, it is used in shipbuilding.

Copper's toxicity makes it effective in preventing bacterial growth; copper doorknobs in hospitals prevent spread of disease. Alas the Canadian penny or "copper" is no longer made of copper. Since 2000, pennies have been made primarily of steel with only 4.5 percent copper to give the coins their copper colour.

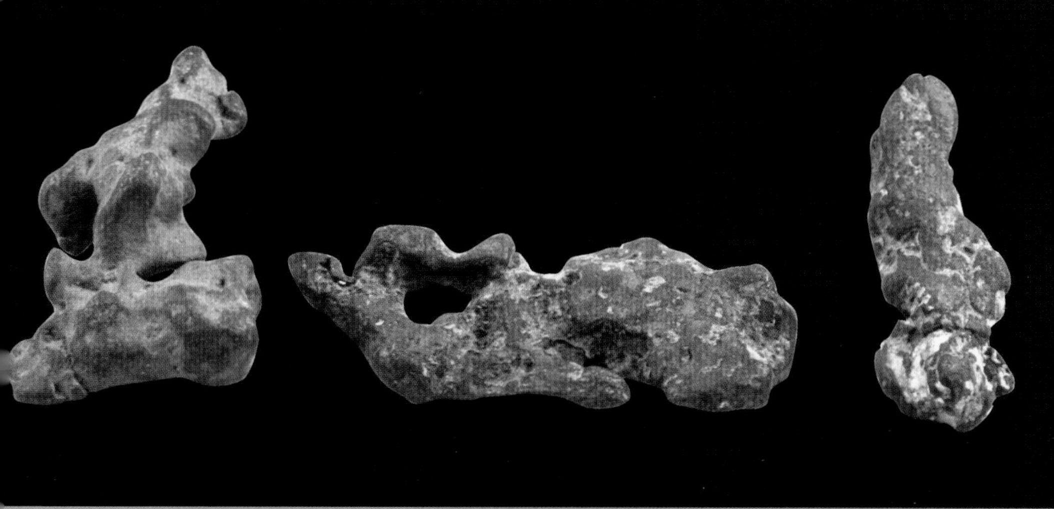

Copper (31376): Waterworn nuggets from the beach. Cape d'Or, Cumberland Co., Nova Scotia. Width of field of view: 13 cm

SILVER: Ag

Silver is considered one of the "precious metals-- because of its scarcity and attractive physical properties. It can occur in nature as "native" silver or combined with other elements such as sulphur (S), chlorine (Cl), oxygen (O), and arsenic (As) making up over two hundred different mineral species. Silver's brilliant white colour, malleability, and resistance to oxidation has made it a valuable ornament as far back as 4000 BCE and as coinage since 800 BCE in all countries between the Indus and the Nile. In pure air, silver will not tarnish or turn black but as our air commonly contains sulphur we are used to the blackening of the metal. It is almost as soft as gold.

Appearance

Colour: Silver-white and often tarnishing to grey or black.
Streak: White.
Lustre and transparency: Shiny, metallic lustre and opaque.
Streak: Brown-black with a greenish tinge.
Habit: Usually massive in lumps, or rounded grains or scales. Rare crystals are octahedral or in dendritic or fern-like aggregates.

Physical Properties

Hardness: Medium (Mohs 2½ – 3).
Density: Heavy (10.1 g/cm^3).
Breakage: Malleable, hence difficult to break. As silver is sectile it can be cut with a knife. There is no cleavage and the hackly fracture described results in a jagged irregular surface like that of broken iron.
Test: Colour, tarnish, sectility, and hackly fracture are the most useful identifying features.

Silver (30930): Wire silver. Highland Bell Mine, Beaverdell, Similkameen Division, Yale District, British Columbia. Width of field of view: 9 cm

Similar Minerals

Galena (p. 56) is greyer in colour and streak and very brittle.

Molybdenite (p. 60) is greyer in colour and streak and has a perfect cleavage.

Silver (32029): Dendritic or fern-like habit.
Cobalt, Coleman Co., Timiskaming District, Ontario. Width of specimen: 8 cm

Silver (53663): Massive silver called the "silver pavement" by miners.
Wettlaufer Mine, South Lorrain Tp., Timiskaming District, Ontario.
Width of specimen: 15 cm

Stibnite (p. 62) is only slightly sectile, has a good cleavage, and is more grey in colour. Arsenopyrite (p. 54) and cobaltite (p. 64) are much harder and more brittle than silver.

Occurrence

Silver is found in the late stages of crystallization of a magma in the presence of hot waters (hydrothermal deposits).

Canada's Best Localities: Cobalt, Timiskaming County, Ontario has produced large sheets, "silver pavement", and wires up to 6 cm; Camsell River, MacKenzie District, Northwest Territories; Highland Bell Mine, Yale District, British Columbia.

Other Localities: Kongsberg, Buskerud, Norway, is the most famous locality for "horn" and "wire" silver; Keweenaw Peninsula, Houghton County, Michigan, USA; Chihuahua, Batopilas District, Mexico.

Interesting Facts

The mineral name is derived from Old English, *seolfor*, c. 450 – 1100. The chemical symbol, Ag, is from Latin, *argentum*, meaning silver, white money. The Latin is from the Proto-Indo-European root, argent, meaning the shining white metal. Refined silver is often sold as Sterling silver which is 92.5 percent Ag (925 fine) and the remainder is mostly copper.

Most of the silver mined today is for the photographic industry, although with film disappearing this use will be lessened. Silver has the highest known electrical and thermal conductivity of any metal, and thus is used in printed electrical circuits.

Silver (40418): Reticulated silver in quartz.
Terra Mine, Camsell River, Northwest Territories. Width of specimen: 14 cm

Silver (39443):
Wire silver on calcite.
Silver Islet, Sibley Tp.,
Sudbury District, Ontario.
Width of field of view: 4 cm

PLATINUM: Pt

Platinum was first found early in the 16th century in South America. Mixed with gold in the eroded placer gravel deposits of the Pinto River, Chocó region of Colombia, South America, it was cursed as an undesirable metal. The grains were removed and thrown back into the river. Now platinum is our most valuable metal, fetching approximately twice the value of gold. It is one of the "noble metals," so called because it is chemically non-reactive.

Appearance

Colour: Silver white to steel grey. It does not tarnish.
Streak: Whitish grey.
Lustre and transparency: Shiny, metallic lustre and opaque.
Habit: Often as grains or nuggets. Crystals are rare.

Physical Properties

Hardness: Hard for a metal (Mohs 4 – 4½).
Density: Very heavy (21.5 g/cm^3).
Breakage: Malleable and ductile with no cleavage. The fracture is hackly.
Test: Colour and malleability.

Similar Minerals

Silver (p. 22) is softer (Mohs 2½ – 3), lighter in density (10.1 g/cm^3), and is often tarnished to a grey or black colour.

Occurrence

Platinum occurs as blebs or rounded inclusions in basic igneous rocks or as nuggets in the gravels of placer deposits associated with igneous rocks.
Canada's Best Locality: Tulameen River, Similkameen District, British Columbia.
Other Localities: Salmon River, Bethel, Alaska, USA; Pinto River, Cauca, Colombia; Permskaya Oblast, Russia.

Interesting Facts

The name is derived from the Spanish *platina* meaning "little silver" due to its resemblance to silver.

Platinum's uses are directly related to what it won't do, that is, its extreme non-reactive nature. It resists acids, making it an ideal substance in the production of nitric acid, an essential ingredient in fertilizer. As a crucible it withstands the extreme heat (melting point 1,769ºC) needed in the manufacturing of ruby laser rods. The majority of platinum is used in automobile catalytic converters, to change noxious gases into harmless carbon dioxide and water. In the cancer treatment drug, cisplatin, two atoms of chlorine and two molecules of ammonia are bonded to a platinum atom. This drug attacks and kills cancer cells, but how it works remains a mystery.

Platinum (OJ 1261): Nugget (91 g) from Suchowisimsk, Urals, Russia. Length of specimen: 4 cm

BISMUTH: Bi

Bismuth occurs both in the native state as bismuth metal and in combination with other elements in more complex minerals. In nature bismuth is quite rare but synthetic, skeletal crystal groups can be readily purchased.

Although classified as a metal it has some rather unique properties uncharacteristic among metals: It expands upon freezing; it is a poor conductor of electricity and heat; it is brittle and not malleable. It is a very heavy metal yet less toxic than other heavy metals such as lead, thallium, and antimony, and it is the most diamagnetic of all metals. Diamagnets are repelled by a magnetic field, and, if this repulsion is strong enough, levitation will occur.

Appearance

Colour: Silver-white with a reddish tinge. Sometimes described as the colour of horse flesh, which may not be a terribly useful diagnostic. Tarnishing darkens the colour and iridescent hues develop.

Streak: Silver-white.

Lustre and transparency: Dull metallic lustre and opaque.

Habit: Often granular and sometimes showing a tree-like growth pattern.

Physical Properties

Hardness: Soft (Mohs 2 – 2½).

Density: Heavy (9.7 g/cm^3).

Breakage: Brittle with a distinct cleavage which gives a stepped appearance; sectile.

Test: Colour, hardness, sectile quality, and stepped cleavage help identify bismuth.

Similar Minerals

Silver (p. 22) is malleable and is silver-white while bismuth is sectile and silver-white with a distinct reddish tinge.

Galena (p. 56) is silver-grey, brittle with perfect cleavage while bismuth is silver-white, sectile with no cleavage.

Occurrence

Bismuth is rare. It occurs in veins formed by hot thermal solutions. It is often associated with ore minerals such as silver, cobaltite and galena.

Canada's Best Localities: Coarse cleavable masses to several centimetres have been found in the Cobalt area, Timiskaming County, Ontario; Mount Pleasant Mine, Charlotte County, New Brunswick; Camsell River, MacKenzie District, Northwest Territories.

Other Localities: Llallagua, Potosi Department, Bolivia; Saxony, Germany; Cornwall, England.

Interesting Facts

In his 15th century writings, German alchemist Basilius Valentinus referred to this metal as *wisimut*, from German *wis mat,* or white metal. Georgius Agricola, the 15th century German scientist considered the father of mineralogy, Latinized the term to *bismutum*. For a long period prior to Agricola bismuth was not recognized as a metal distinct from lead, antimony, or tin. Earlier, in the Middle Ages, it was believed bismuth was an early stage in the formation of silver.

Bismuth has two special properties that lead to many important uses: low melting point and volume expansion upon solidification of its liquid melt. As an ingredient in low-melting alloy it is used in fire detection equipment. Most substances contract upon solidification; exceptions are water to ice and liquid bismuth to solid bismuth. This property makes bismuth useful in forming sharp, clean casts. In the chemical industry bismuth is an effective agent in the manufacture of acrylic fibres, paints, and plastic. For centuries bismuth has been used in the treatment of indigestion, diarrhea, and wounds. It is also used as a replacement for lead in sinkers used in fishing, and in gunshot.

Bismuth (31564): Distinct cleavage gives a stepped appearance.
Kerr Lake Mine, Cobalt, Coleman Tp., Timiskaming District, Ontario. Width of field of view: 10 cm

GRAPHITE: Carbon: C

Carbon occurs as minerals with a variety of crystal structures – graphite and diamond being the most noteworthy. These two polymorphs (things having the same chemical composition but a different crystal structure) are an excellent example of how chemical bonding affects the physical properties of a compound. In 1798 the famous German mineralogist Abraham Gottlob Werner coined the German word, *graphit*, meaning "black lead" to name this mineral. For most of us the lead pencil is our first acquaintance with graphite but now carbon derived from graphite has many more exotic uses.

Appearance

Colour: Black to steel-grey.
Streak: Black.
Lustre and transparency: Sometimes has a metallic lustre but it is often dull or earthy and opaque.
Streak: Brown-black with a greenish tinge.
Habit: Commonly found as foliated or earthy masses. Rarely found crystals are tabular.

Physical Properties

Hardness: Very soft (Mohs 1 – 2).
Density: Light (2.2 g/cm^3).
Breakage: Sectile, flexible, but not elastic, breaks easily with a perfect cleavage face.
Test: Very soft and feels greasy.

Similar Minerals

Molybdenite (p. 60) is shinier in lustre and has a greenish tint in the streak. Stibnite (p. 62) is more silvery in colour than graphite and it has a lead-grey streak.

Occurrence

Graphite is found in metamorphosed sedimentary rocks that contained organic material or carbonate minerals. The metamorphic rock can be slate, schist, or gneiss.

Canada's Best Localities: Castor Lake, Joly Township, Labelle County, Québec; Pointe-au-Chêne, Grenville Township, Argenteuil County, Québec; Buckingham Township, Papineau County, Québec; Black Donald Mines, Broughton Township, Renfrew County, Ontario; Kimmirut, Baffin Island, Nunavut.
Other Localities: Stirling Hill Mine, Sussex County, New Jersey, USA; Hunan, China.

Graphite (31436): Foliated and massive with a sub-metallic lustre.
Grenville Augmentation Mines, Pointe-au-Chêne, Grenville Tp., Argenteuil Co., Québec.
Width of specimen: 7 cm

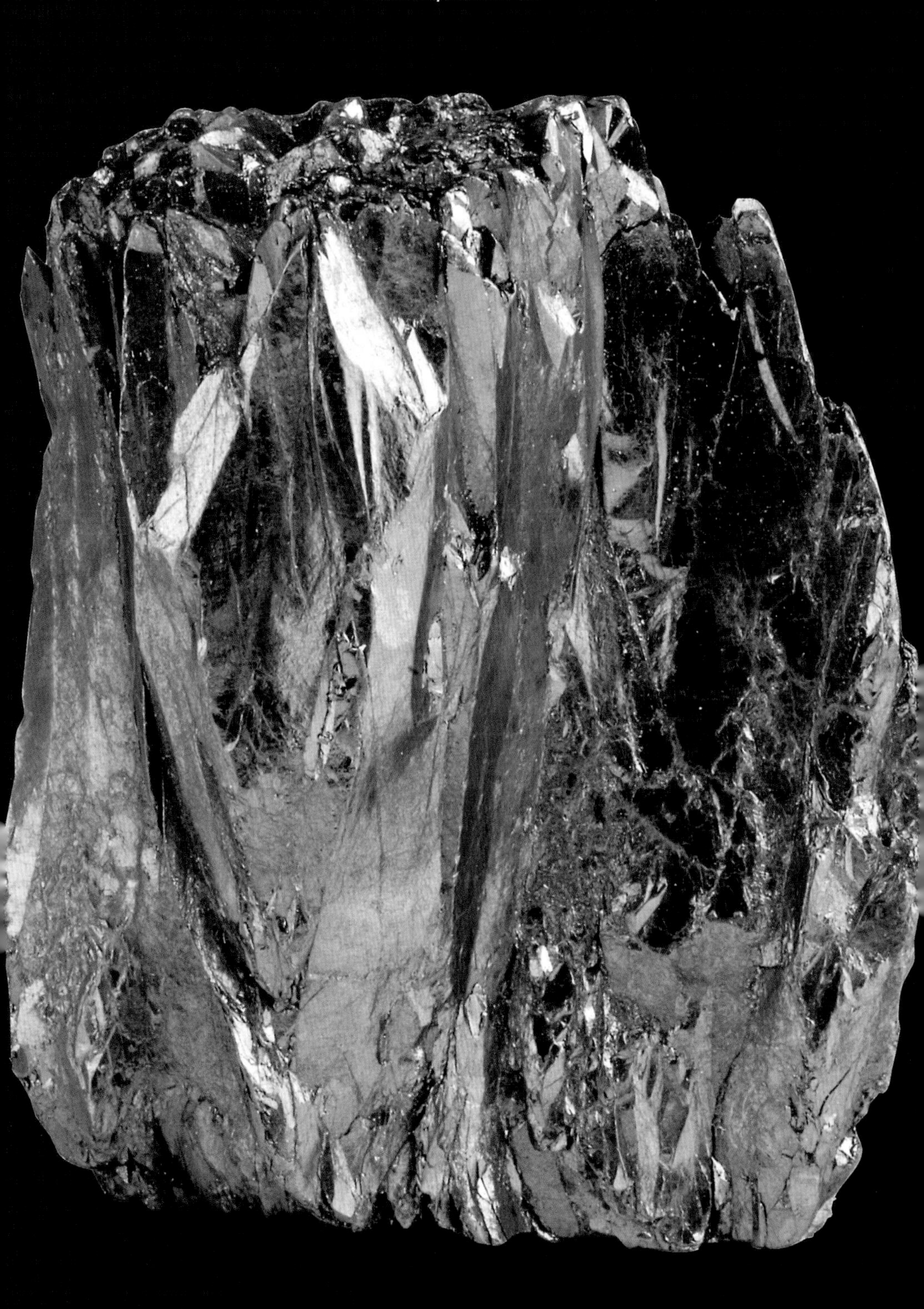

Interesting Facts

The mineral name is from the Greek word, *graphein*, meaning "write," so called as it was used in pencils.

The excellent cleavage of graphite makes it useful in dry lubricants. Synthetic forms of graphite (pyrolytic carbon) are extremely strong and heat resistant (up to 3000°C). These synthetic forms find uses in missile nose cones, batteries, foundry crucibles, brushes for electric motors, and nuclear reactor cores. Synthetic carbon (carbon fibre) is also used to reinforce plastics in fishing rods, racing bicycles, squash racquets, and golf clubs. Interesting medical applications arise because pyrolytic carbon discourages blood clotting, hence its usage in artificial hearts and heart valves. It is also used in knuckle and spinal implants.

Graphite (46671): Fibrous.
Soper River, Kimmirut, Baffin Island, Nunavut. Width of specimen: 15 cm

Graphite (49682): Platy crystals on feldspar.
Old Graphite Mine, Clemons, Washington Co., New York. Width of specimen: 15 cm

CHALCOCITE: Copper sulphide: Cu_2S

Chalcocite is an important copper ore and with 80 percent copper by weight it is one of the richest. Unfortunately it is not a major ore as it is quite rare. Fine crystals of chalcocite are rare and much sought after by collectors. As it has such a high metal content it tends to be sectile like a metal.

Appearance

Colour: Dark grey when fresh to black when altered.
Streak: Shiny black to lead-grey.
Lustre and transparency: Metallic lustre and opaque.
Habit: Most commonly massive but sometimes tabular, hexagonal-shaped crystals are found.

Physical Properties

Hardness: Moderate (Mohs 2½ – 3).
Density: Heavy (5.7 g/cm^3).
Breakage: Almost brittle and somewhat sectile with conchoidal fracture.
Test: Colour, hardness, and sectile nature are important.

Similar Minerals

Covellite (p. 36) often has a purple tarnish, is more brittle, and has a cleavage.

Chalcocite (31659): Massive, dull metallic lustre. Old Borron Location, Chubb Lake area, Gould Tp., Algoma District, Ontario. Width of specimen: 9 cm

Occurrence

Chalcocite is found in copper deposits that have been enriched with warm water solutions (hydrothermal). In many occurrences it forms as a replacement of other minerals (termed secondary) such as chalcopyrite, and covellite.

Canada's Best Localities: Mount Pleasant, Charlotte County, New Brunswick; Normandie Mine, Mégantic County, Québec; Coppercorp Mine, Algoma District, Ontario; Kidd Creek Mine, near Timmins, Cochrane District, Ontario; Great Bear Lake, Northwest Territories.
Other Localities: Butte, Montana, USA; Hartford County, Connecticut, USA; Messina District, Transvaal, South Africa; Cornwall, England.

Interesting Facts

The name chalcocite comes from the Greek *chalkos*, meaning copper, and reflects the mineral's composition. It was referred to as 'redruthite' by Agricola, one of the first mineral experts, in 1546.

Chalcocite is often a late-stage mineral, replacing other minerals as a pseudomorph.

Chalcocite (31625): Complex crystals with prismatic and tabular habits. Metallic lustre. Cornwall, England. Length of specimen: 11 cm

COVELLITE: Copper sulphide: CuS

Covellite, like chalcopyrite, is an ore of copper, but covellite is much less common than chalcopyrite. It has a beautiful, intense purple-blue colour that makes it popular with jewellery makers. Covellite is often associated with other copper minerals. Prospectors must learn to tell covellite and chalcopyrite apart as they reflect different grades of copper content in the ore.

Appearance

Colour: Dark blue to black, often with a purple tarnish.

Streak: Black with a detectable shine or metallic appearance.

Lustre and transparency: Metallic lustre and opaque.

Habit: Rarely occurs as crystals. Crystals are flat or tabular, sometimes with a hexagonal shape but more commonly just tabular with hexagonal striations on the flat face. It is usually massive, sometimes coating other copper sulphide minerals.

Physical Properties

Hardness: Soft (Mohs 1 – 1½).

Density: Medium (4.7 g/cm^3).

Breakage: Brittle with a distinct cleavage. Cleavage fragments are slightly flexible.

Test: Colour, lustre, and hardness are important.

Similar Minerals

Bornite (p. 38), when fresh, tends to be more bronze-coloured. Tarnished specimens of bornite could easily be confused with covellite.

Chalcocite (p. 34) tends to be more grey to black, sectile not brittle, with no purple or blue tarnish.

Occurrence

It is found in copper enriched zones of copper deposits. These enriched zones are a result of sulphur being removed or leached out by warm circulating water. It is often associated with chalcopyrite, chalcocite, bornite, and pyrite.

Canada's Best Localities: Definitely a rare mineral in Canada, Afton Mine, Yale District, British Columbia.

Other Localities: Butte, Silver Bow County, Colorado, USA; Rio Grand County, Colorado, USA; Kennecotte, Alaska, USA; Alghero, Sardinia, Italy.

Interesting Facts

The mineral named for the Italian mineralogist, Niccolo Covelli (1790–1829), who first described the mineral from an occurrence at Mount Vesuvius, Italy. The uses of copper are listed with chalcopyrite (p. 42).

Covellite (57362): Hexagonal, platy crystals.
Leonard Mine, Butte, Silver Bow Co., Montana. Width of specimen: 9 cm

Covellite (80341): Massive with purple iridescence.
East Grey Rock Mine, Butte, Silver Bow Co., Montana. Width of specimen: 8 cm

BORNITE: Copper iron sulphide: Cu_5FeS_4

Bornite is an important ore of copper that has been mined for hundreds of years. It was first described as a mineral in the late 18th century. One of the most characteristic properties of bornite is its colour.

Appearance

Colour: Freshly broken surface is copper-red to bronze to golden brown. Bornite quickly tarnishes to an iridescent blue to purple, then black.

Streak: Greyish black.

Lustre and transparency: Dull, metallic lustre and opaque.

Habit: Bornite is usually massive or granular without crystal faces. Rarely one finds a cubic-like crystal.

Physical Properties

Hardness: Moderate (Mohs 3).

Density: Heavy (5.1 g/cm^3).

Breakage: Brittle with no even cleavage faces.

Test: Colour and streak are the important properties.

Similar Minerals

Chalcopyrite (p. 42) is brass-yellow in colour, does not tarnish as brightly, and has a greenish black streak.

Covellite (p. 36) is black on the fresh surface while bornite is bronze or copper-red. Both minerals are brittle but only covellite has a cleavage.

Chalcocite (p. 34) is grey when freshly broken and it tends to be more sectile.

Occurrence

Bornite is relatively common in rich copper ore deposits but it is much less common than chalcopyrite. It is often found with chalcopyrite, pyrite, and pyrrhotite in massive, heavy ore samples.

Canada's Best Localities: Harvey Hill Mine, Mégantic County, Québec; Cupra Mine, Wolfe County, Québec; Kidd Creek Mine, Cochrane District, Ontario; Gowganda, Timiskaming District, Ontario; Lornex Mine, Kamloops District, British Columbia.

Other Localities: Butte, Montana, USA; Pinal County, Arizona, USA; Messina District, Transvaal, South Africa; Kipushi Mine, Zaire; Dzhezkazgan, Kazakhstan.

Bornite (45788): Tarnished (purple "peacock" ore) with fresh bronze coloured bornite. Kirkland Lake area, Teck Tp., Timiskaming District, Ontario. Length of specimen: 10 cm

Interesting Facts

Bornite is named after a famous Austrian mineralogist, Ignaz Edler von Born (1742–1791). Born invented an amalgamation process for removing gold and silver from copper ores without the usual, costly melting of the ore.

Bornite is often referred to as "peacock ore" by miners and prospectors because of its distinctive purple-blue tarnish. In Cornwall, England, miners described this mineral as "horseflesh ore" comparing the colour of freshly broken bornite to the raw meat of the animal. Another colour term commonly used in textbooks is "pinchbeck," a copper-zinc alloy used to imitate gold in inexpensive jewellery.

Bornite (80354): Rare crystals.
Leonard Mine, Butte, Silver Bow Co., Montana. Width of field of view: 7 cm

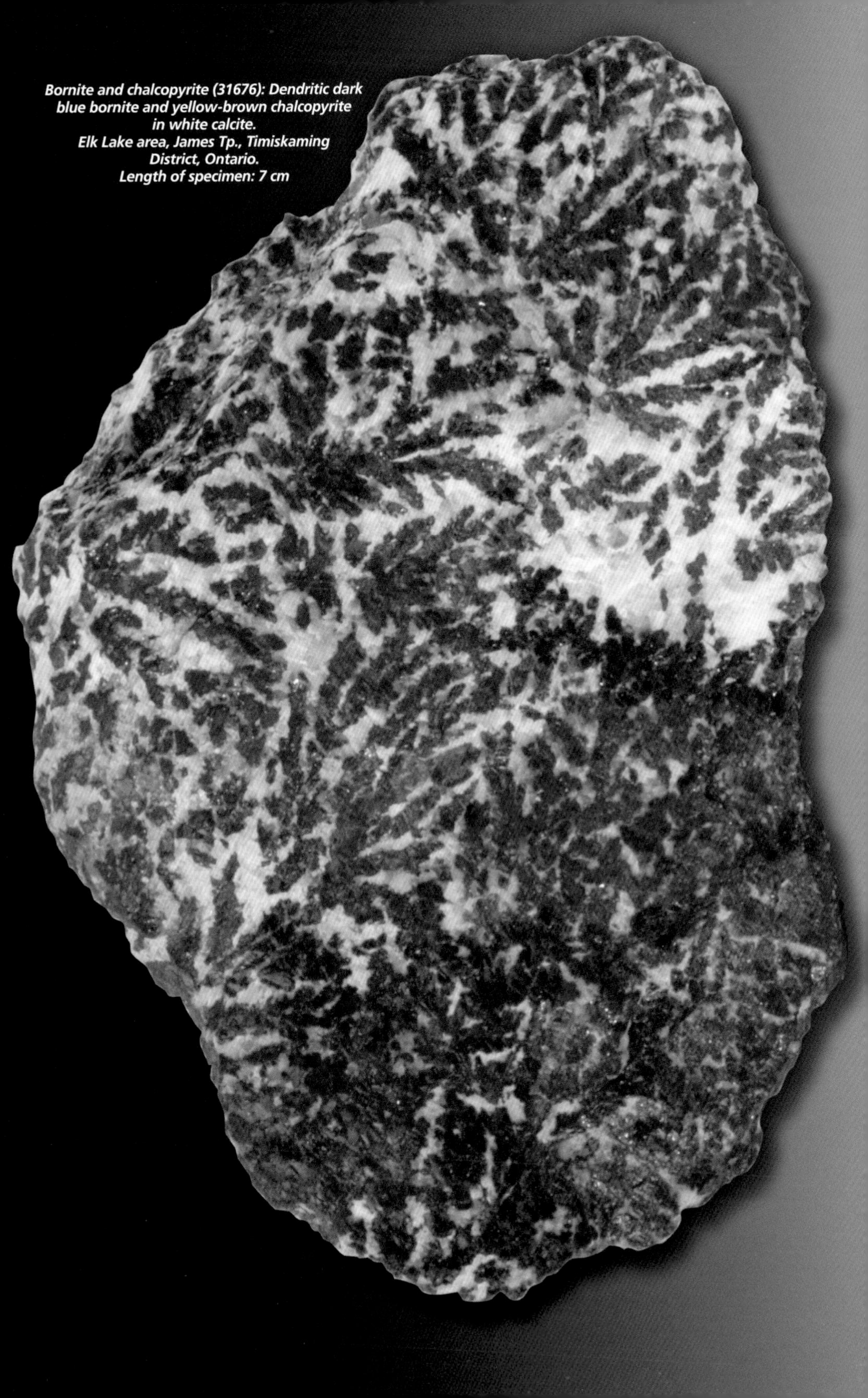

Bornite and chalcopyrite (31676): Dendritic dark blue bornite and yellow-brown chalcopyrite in white calcite.
Elk Lake area, James Tp., Timiskaming District, Ontario.
Length of specimen: 7 cm

CHALCOPYRITE: Copper iron sulphide: $CuFeS_2$

Chalcopyrite is the most common ore of copper, particularly in Canada. It and pyrite are known as fool's gold and of the two, chalcopyrite, with its bright yellow colour, is the more likely to be confused with gold.

Chalcopyrite is often found in association with bornite. Mineralogists termed them "peacock ores" because the tarnish of the two minerals gives shades of green, blue, and purple, reminiscent of the colours in a peacock's tail feathers.

Of the four main ore minerals of copper, chalcopyrite has the least copper by weight (35%), with bornite next (63%), then covellite (66%), and finally the richest, chalcocite (80%).

Appearance

Colour: Brassy yellow, sometimes tarnished with slightly iridescent greenish and bluish colours.
Streak: Greenish black.
Lustre and transparency: Bright metallic lustre and opaque.
Habit: Usually massive but sometimes found as tetrahedral-shaped crystals.

Physical Properties

Hardness: Moderate (Mohs 3½ – 4).
Density: Medium (4.2 g/cm³).
Breakage: Brittle with an uneven fracture and no consistent cleavage faces.
Test: Streak is the most helpful property.

Similar Minerals

Pyrite (p. 46) is harder and less yellow in colour. Pyrite often occurs as cubic crystals while chalcopyrite seldom occurs as crystals.

Gold (p. 18) is malleable, softer, not as brittle, and has a golden-yellow streak. Gold is a more yellow gold, not as brassy as chalcopyrite.

Bornite (p. 36) is bronze-coloured and has a greyish black streak.

Occurrence

Chalcopyrite is the most commonly occurring copper mineral. It is found in volcanic rocks often concentrated by hydrothermal solutions. It is often associated with one or more of pyrite, pyrrhotite, pentlandite, sphalerite, and magnetite.

Canada's Best Localities: Temagami Mine, Nipissing District, Ontario; Frood Mine, Sudbury District, Ontario; Bicroft Mine, Cardiff Township, Haliburton County, Ontario; Courtenay, Comox District, British Columbia.

Other Localities: St. Lawrence County, New York, USA; Bisbee, Cochise County, Arizona, USA; Grant County, New Mexico, USA; Chester County, Pennsylvania, USA; Zacatecas, Mexico; Cornwall, England; Herja, Romania; Ugo, Japan; Yorke Peninsula, South Australia, Australia.

Chalcopyrite and pentlandite (45786): Massive, metallic lustre, chalcopyrite is yellow, pentlandite is bronze colour.
Denison Tp., Sudbury District, Ontario.
Width of field of view: 17 cm

Chalcopyrite (59382): Rounded tetrahedra on calcite.
French Creek Mines, Knauertown,
Chester Co., Pennsylvania.
Width of field of view: 6 cm

Interesting Facts

The mineral name is derived in two parts: *chalkos*, Greek meaning copper, and *pyrite*, which it resembles. It is believed that chalcopyrite was first smelted in the Middle Ages. In this process of heating or smelting the sulphur is driven off and the iron is extracted using sand, which forms a slag, or glassy iron silicate, and leaving almost pure copper metal.

Copper finds many uses in electrical wiring, plumbing, roofing, and alloys of bronze (copper, tin, and sometimes lead) and brass (copper and zinc). Less common alloys are "German silver" (a mixture of copper, zinc, and silver) which is used in musical instruments and "aluminum bronze" (a mixture of copper and aluminum), which is useful in jewellery because of its golden colour.

PYRITE: Iron sulphide: FeS_2

Pyrite is a mineral that catches the eye because of its seductive, golden, metallic lustre. Despite its beautiful appearance, it is a very common mineral and is far less valuable than gold. In the gold rush days, pyrite was nicknamed "fool's gold." Pyrite crystals are still of great interest to collectors if they have large crystal faces with distinct crystal forms. Some crystals form perfect cubes or irregular pyramids, while others cluster together in aggregates. Under special fossilization conditions, pyrite can replace marine shells, forming stunning, golden fossils. For identification, use the streak and hardness tests.

Appearance

Colour: Pale brass-yellow, but sometimes tarnished to brown.

Streak: Brown-black with a greenish tinge.

Lustre and transparency: Shiny, metallic lustre and opaque.

Pyrite (46954): Cubic form in white nepheline. Demix & Poudrette Quarries, Mt. St.-Hilaire, Rouville Co., Québec. Width of field of view: 5 cm

Habit: Pyrite can be massive, without crystal faces. Often good crystals have the form of a cube, an octahedron, or a pyritohedron.

Physical Properties

Hardness: Hard (Mohs 6).

Density: Heavy (5.0 g/cm^3).

Breakage: Brittle with no even cleavage faces.

Test: Makes sparks when struck with pyrite or an iron object.

Pyrite (56682): Octahedral crystals. Santiago de la Libertad Mine, Quiruvilca, Santiago de Chuco Province, Peru. Width of field of view: 8 cm

Similar Minerals

Gold (p. 18) is softer, not as brittle, and has a golden-yellow streak.
Pyrite has a silvery tinge in comparison to a true gold colour, and a darker streak.

Chalcopyrite (p. 42) may resemble tarnished pyrite but pyrite is harder.

Marcasite (p. 48) is very similar to pyrite but has a less intense, yellow colour.

Occurrence

Pyrite is found in all types of rock and is often associated with ore minerals such as magnetite. Pyrite is the most common mineral in the sulphur-containing minerals called sulphides.

Canada's Best Localities: Crystals up to 25 cm have been found at Snow Lake, Manitoba. Pyrite cubes containing tourmaline (p. 188) inclusions come from Marmora, Hastings County, Ontario; Nunavut.

Other Localities: Oddly shaped and exceptional crystals can be found at the following locations: joined, perfect cubes from Navajun, Logrono, Spain; octahedral crystals from Huanzala, Huanuco, Peru; and complex pyritohedral crystals from Rio Marina on the island of Elba.

Pyrite (46449): Brachipod shell replaced by pyrite in black shale. Mt. St.- Bruno area, Chambly Co., Québec. Width of field of view: 5 cm

Interesting Facts

The mineral name is derived from the Greek word *pyr*, meaning fire, because it makes sparks when struck. Although it has iron (Fe) in the chemical formula, pyrite is not often mined for iron, but is an important source of sulphur (S) used in the production of sulphuric acid. Ironically, pyrite is often disappointingly confused with gold, but in some occurrences pyrite contains small amounts of gold. This source of gold can be recovered and be very profitable.

Pyrite (83605): Complex crystal with pyritohedron (pentagonal dodecahedron) as the dominant form with cube (rectangular shape) on some crystals. Nanisivik Mine, Nanisivik, Baffin Island, Nunavut. Width of field of view: 9.5 cm

MARCASITE: Iron sulphide: FeS_2

Marcasite is a polymorph (more than one form) of pyrite, meaning it has the same chemical composition as pyrite but a different arrangement of atoms in its atomic structure. In fact it is one of the best known polymorphs, as both minerals are quite common. Marcasite forms in lower temperature environments than pyrite and it can be found as recently formed masses in lake bottoms.

Not surprisingly marcasite and pyrite strongly resemble each other but because they have different crystal structures one can note distinct differences in crystal habits and minor differences in their physical properties. Marcasite is noted for its replacement of other minerals such as pyrite, gypsum, and fluorite in pseudomorphs. Iron and sulphur atoms within the marcasite structure replace the atoms of the original mineral, thus creating the pseudomorph.

Appearance

Colour: Pale brass-yellow, becoming slightly darker in colour with exposure.

Streak: Greyish black.

Lustre and transparency: Metallic lustre and opaque.

Habit: Often as crystals in a variety of shapes: prismatic and tabular crystals are common, often having curved faces; twinned aggregates can be spear-shaped or "cock's comb" when it resembles the crest on a rooster's head. It may also be massive in stalactitic or rounded habits.

Physical Properties

Hardness: Hard (Mohs 6 – 6½).

Density: Medium (4.9 g/cm^3).

Breakage: Brittle with a distinct cleavage and a conchoidal fracture on the non-cleavage planes.

Test: Colour and hardness are important. Deterioration (oxidation in air) of specimen to a white powder is associated with a sulphur smell.

Similar Minerals

Pyrite (p. 46) has no cleavage and it is usually a little more intensely yellow than marcasite. Marcasite breaks down more readily than pyrite; thus it may have yellowish sulphate crusts on it.

Chalcopyrite (p. 42) is softer (Mohs 3½ – 4) than marcasite.

Occurrence

Marcasite is found most often in sedimentary environments as crystals or masses in limestone, shale, or clay. In igneous environments it is found where low-temperature hydrothermal solutions have deposited it with sphalerite and galena.

Canada's Best Localities: Inuvik, Northwest Territories; George McLeod Mine, Wawa, Chabanel Township, Algoma District, Ontario.

Other Localities: Shullsburg, Wisconsin, USA; Rosiclare, Hardin County, Illinois, USA; Joplin

District, Missouri, USA; Ross County, Ohio, USA; Picher, Oklahoma, USA; Misburg, Saxony, Germany; Vintirov, Bohemia, Czech Republic; Cavnic, Romania.

Interesting Facts

The mineral name is derived from *marcasita* and refers to all minerals resembling pyrite. Further confusion in the marcasite/pyrite story is that "marcasite jewellery" is not marcasite but pyrite.

Marcasite readily breaks down to a fine yellowish white powder making it difficult to keep specimens in a collection for a long period of time. Deterioration is enhanced in fine-grained specimens or in humid conditions. This oxidation, sometimes called "pyrite disease," forms sulphuric acid which will attack the label and box it sits in.

Marcasite (36329): Coxcomb habit. Helen Mine, Wawa, Chabanel Tp., Algoma District, Ontario. Width of field of view: 7 cm

PYRRHOTITE: Iron sulphide: $Fe_{1-x}S$

Pyrrhotite is a common mineral in copper and nickel ore deposits. It is one of several brassy, yellow sulphide minerals including pyrite, marcasite, pentlandite, and chalcopyrite. The formula may strike you as odd as there is a variable amount of iron. To date five different crystal-structure varieties have been found. The more iron-deficient pyrrhotite is magnetic. This may be contrary to what you would think as in general the more iron-rich compounds are more magnetic: The reason for this phenomenon is that the iron deficiency leads to freed electrons that impart magnetism. Crystals are rare and thus desirable to collectors.

Appearance

Colour: Bronze-yellow on a fresh surface but it readily tarnishes to brown, often with iridescence.

Streak: Black to dark grey.

Lustre and transparency: Metallic lustre and opaque.

Habit: Often massive but hexagonal, prismatic crystals can be found.

Physical Properties

Hardness: Medium (Mohs 4).

Density: Heavy (4.6 g/cm^3).

Breakage: Brittle with no distinct cleavage and an uneven fracture.

Test: Usually magnetic but with varying intensity. Best measured on small fragments.

Pyrrhotite (30118): Stubby, hexagonal prismatic crystals on clear quartz crystals. Bluebell Mine, Riondel, Kootenay District, British Columbia. Width of field of view: 8 cm

Pyrrhotite (31933): Massive, granular with indication of basal pinacoid of prismatic crystals. Strathcona Mine, Sudbury Basin, Levack Tp., Sudbury District, Ontario. Width of field of view: 14 cm

Similar Minerals

Pyrite (p. 46) is harder (Mohs 6), non-magnetic, and lighter in colour with a slight greenish tint to the streak.

Chalcopyrite (p. 42) is brassy yellow, hence lighter in colour with a slight greenish tint in the streak.

Pentlandite (p. 52) has a bronze-yellow streak and it is not magnetic.

Occurrence

Most commonly pyrrhotite is found in igneous rocks and associated veins. It can be found in association with pyrite, galena, chalcopyrite, and pentlandite.

Canada's Best Localities: Bright crystals to 10 cm were found at Bluebell Mine, Riondel District, British Columbia; Henderson #2 Mine, McKenzie Township, Abitibi County, Québec; Thompson Mine, Manitoba.

Other Localities: Santa Eulalia, Chihuahua, Mexico; Belo Horizonte, Minas Gerais, Brazil; Baia Sprie, Romania; Trepca, Yugoslavia.

Interesting Facts

The mineral name is from Greek, *pyrrotes*, reddish, in reference to the faint tint of red similar to that seen in bronze.

Pyrrhotite has no value as an iron ore because of the environmental problem of dealing with the sulphur emissions on smelting. The formula tells us that there are always fewer iron atoms than sulphur atoms. This chemical imbalance gives pyrrhotite its magnetic properties. Pure FeS does exist in meteorites as its own mineral species, troilite.

Between 1956 and 1991 iron was recovered from the Sudbury ores by processing the slag, mostly iron silicate, resulting from processing the copper-nickel ores.

PENTLANDITE: Iron and nickel sulphide: $(Fe,Ni)_9S_8$

Pentlandite is an important ore mineral of nickel but unfortunately it is difficult to identify. It is one of several bronze-coloured sulphide minerals with no distinguishing crystal form. It is often intimately associated with one of its look-alikes – pyrrhotite, pyrite, and chalcopyrite. It takes practice to identify pentlandite. Before the development of important nickel uses it was considered a nuisance mineral interfering with copper production.

Appearance

Colour: Light bronze-yellow.
Streak: Bronze-yellow.
Lustre and transparency: Metallic lustre and opaque.
Habit: Massive to granular.

Physical Properties

Hardness: Moderate (Mohs 3½ – 4).
Density: Moderately heavy (4.8 g/cm³).
Breakage: Brittle with conchoidal fracture; sometimes the broken surface appears angular due to a parting on twin planes.
Test: Colour and hardness.

Similar Minerals

Chalcopyrite (p. 42) is more brassy, yellow or golden.
Pyrrhotite (p. 50) is very similar in colour yet it is often magnetic.
Pyrite (p. 46) is harder (Mohs 6) and a pale brass-yellow colour.

Occurrence

It is found in massive sulphide ore deposits associated with pyrite, chalcopyrite, and pyrrhotite. This type of sulphide deposit is associated with basic rocks.

Canada's Best Localities: Giant Mascot Mine, Yale District, British Columbia; Thompson Mine, Manitoba; Copper Cliff South Mine, Creighton Mine, and Strathcona Mine, Sudbury District, Ontario.
Other Localities: Outokumpo, Finland; Carr Boyd Nickel Mine, Western Australia.

Interesting Facts

The mineral is named after the Irish naturalist Joseph Barclay Pentland (1797–1873) who first made note of the mineral occurrence in the nickel ores of Sudbury in the 1850s. Initially ores from the Sudbury region were thought to be worthless, as when smelted they produced the dreaded "kupfer-nickel." This name, given by Saxon miners two hundred years before, refers to the hard, light-coloured, hard-to-work, copper-nickel metal refined from their ores. They had hoped for the easy-to-work kupfer, or copper, and in their disappointment they blamed the devil, or "Old Nick" for casting a spell on their ore and rendering it useless.

Stainless steel accounts for the largest use of nickel. In recent years a whole new series of "superalloys" have been developed that are largely nickel and that withstand high temperatures (up to 1100ºC) and resist corrosion. These superalloys are used in aircraft turbines, rocket engines, nuclear reactors, and space vehicles.

Pentlandite (32110): Pale, brassy yellow pentlandite with yellow chalcopyrite with some blue staining. Copper Cliff, Sudbury District, Ontario. Width of field of view: 22 cm

ARSENOPYRITE: Iron arsenic sulphide: FeAsS

Arsenopyrite is the most common arsenic mineral and it is an ore of the element arsenic. It is not mined specifically for arsenic as enough of this element is derived as a by-product in other mining activities.

Arsenopyrite is an attractive mineral with sharp, bright crystals. If heated it will give off toxic arsenic fumes but under normal temperatures it is not dangerous to handle or store in a collection. It is an important mineral to identify, as it is sometimes associated with gold.

Appearance

Colour: Silver-white to steel-grey.

Streak: Black to dark grey.

Lustre and transparency: Shiny, metallic lustre and opaque.

Habit: Often as rhombic, diamond-shaped crystals with striations across the face.

Physical Properties

Hardness: Hard (Mohs 5 – 6).

Density: Heavy (6.1 g/cm^3).

Breakage: Brittle with a distinct cleavage and an uneven fracture on the non-cleavage planes.

Test: Colour and hardness.

Arsenopyrite (41062): Massive arsenopyrite with brassy yellow chalcopyrite. Nigadoo River Mine, Gloucester Co., New Brunswick.
Width of specimen: 7 cm

Similar Minerals

Galena (p. 56) is softer (Mohs 2), and has perfect cleavage in all three planes.

Stibnite (p. 62) is softer (Mohs 2) and has a lead-grey streak. The striations on stibnite crystals are along the length while those on arsenopyrite are across the diamond-shaped face.

Cobaltite (p. 64) has cubic crystals while arsenopyrite has rhombic (diamond-shaped) crystals.

Occurrence

Arsenopyrite is found in high-temperature hydrothermal fracture-filling veins associated with gold and quartz.

Canada's Best Localities: Bluebell Mine, Riondel, Kootenay District, British Columbia has excellent rhombic crystals to 2 cm; Sigma #2 Open Pit, Abitibi County, Québec, has elongate prisms.

Other Localities: Fine crystal groups from Santa Eulalia, Mexico; Panasqueira, Portugal; Trepca, Serbia.

Interesting Facts

The mineral name comes from a contraction of the German for "arsenical pyrites." As early as the time of alchemist Albertus Magnus (13th century) various kinds of "unfinished" metals were recognized. The unfinished metals were minerals that had a metallic lustre but not the other properties of a metal such as malleability. They were given the broad name of *marchasita* and included pyrite, marcasite, arsenopyrite, cobaltite, and stibnite. They were vaguely differentiated by descriptions such as golden, silvery, tinny, and leaden. Although these minerals looked like ores, they always disappointed prospectors and metallurgists because when heated in a furnace they did not yield a desirable metal and gave off noxious fumes.

The element arsenic has been used for over two thousand years. Early uses included poisoning and embalming. Today arsenic is used as a pesticide and herbicide in agriculture. Wood and animal skins are treated with arsenic compounds to protect them from fungi. The element is also used as a colourant in fireworks, as an alloy in superconductor lasers and in LED (light-emitting device) lights. Small amounts of arsenic are used in anti-parasitic drugs and in the treatment of certain types of leukemia.

Arsenopyrite (31210): Diamond-shaped, striated crystals of arsenopyrite on rounded, grey galena and surrounded by white calcite.
Bluebell Mine, Riondel, Kootenay District, British Columbia.
Width of field of view: 7 cm

GALENA: Lead sulphide: PbS

Galena can be found as spectacular crystals with shiny, often stepped faces. Crystals are characteristically perfect cubes and the cleavage is also cubic. The arrangement of atoms in the galena crystal structure is the same as that of halite (salt). The stepped faces are a good demonstration of crystal growth by addition of unit cells. The heft or weightiness is a memorable physical property and it makes one think of lead. Galena is the main ore of lead.

Appearance

Colour: Silver-grey to dull lead colour.
Streak: Lead grey.
Lustre and transparency: Bright, metallic lustre and opaque.
Habit: Galena is usually found as cubic crystals. Sometimes there are octahedral faces.

Galena (32421): Galena crystals with large, triangular, octahedral faces modified by small, square cube face. Note the small resinous, orange sphalerite crystals on a calcite base. Joplin, Jasper Co., Missouri. Width of field of view: 5 cm

Galena (85368): Cubic galena crystals with a small triangular face (octahedron) at each corner Polaris Mine, Little Cornwallis Island, Nunavut. Width of field of view: 12 cm

Physical Properties

Hardness: Medium (Mohs 2½).
Density: Heavy (7.6 g/cm^3).
Breakage: Brittle with perfect cubic cleavage faces.
Test: Streak is important and easy, perfect cleavage.

Similar Minerals

Stibnite (p. 62) is softer, not as brittle, and has a grey-black streak. Stibnite is lighter in weight than galena.

Molybdenite (p. 60) is platy, often hexagonal, and flexible with only one good cleavage.

Occurrence

Galena is found in hydrothermal veins, often associated with sphalerite, pyrite, and chalcopyrite.

Canada's Best Localities: Bluebell Mine, Kootenay District, British Columbia; Sa Dena Hess Mine, Watson Lake, Yukon Territory; Polaris Mine, Little Cornwallis Island, Northwest Territories; Nanisivik Mine, Baffin Island, Nunavut; Dundas Quarry, Flamborough Township, Wentworth County, Ontario; Mount Pleasant, Charlotte County, New Brunswick.

Other Localities: The Tri-State District, including parts of Kansas, Missouri, and Oklahoma, USA, has produced spectacular groups; Rossie, New York, USA; Naica, Chihuahua, Mexico; Bleiberg, Austria; Madan, Bulgaria; Freiberg, Saxony, Germany; Brod and Trepca, Yugoslavia.

Interesting Facts

The mineral name is directly from the Latin, *galena*, meaning lead ore.

Galena is by far the most important ore of lead. Lead is used for radiation protection and in the past it was an important additive in paints and gasoline. It is used in fine, crystal glass to increase the refractive index giving it that wonderful play of colour caused by the refraction of light. Galena can take silver into its atomic structure, thus it is also an important ore of silver. In the 1940s and '50s children and adults alike enjoyed crystal-radio sets and it was galena crystals that served as the semiconductor. This discovery led to a huge industry in electronics.

Galena (32619): Broken galena showing perfect, cubic cleavage planes with white calcite. Kingdon Mine, Galetta, Fitzroy Tp., Carleton Co., Ontario. Width of field of view: 10 cm

MOLYBDENITE: Molybdenum sulphide: MoS_2

Until late in the 18th century this mineral was confused with lead, galena, and graphite. It's still easy to confuse these four minerals, but molybdenite is very bright, shiny, and platy. Beautiful hexagonal plates and roses have been found.

Appearance

Colour: Shiny lead-grey colour with a bluish tint.

Streak: Grey with a greenish tint.

Lustre and transparency: Shiny, metallic lustre and opaque.

Habit: Usually foliated or bent, thin plates, often with a hexagonal outline. Sometimes massive.

Molybdenite (33459): Metallic, steel-grey with perfect cleavage. Maniwaki area, Egan Tp., Gatineau Co., Québec. Width of specimen: 8 cm

Physical Properties

Hardness: Soft (Mohs 1 – 1½), feels greasy.

Density: Heavy (4.7 g/cm^3).

Breakage: Flexible but not elastic (bends and stays bent). One perfect cleavage face parallel to the sheets.

Test: The streak colour, perfect cleavage, and flexible plates are important physical properties.

Similar Minerals

Graphite (p. 30) is similar in hardness and cleavage but has a grey-black streak. The colour of graphite is more grey-black while molybdenite has a bluish tint and is more reflective or shiny.

Galena (p. 56) has three perfect cleavages while molybdenite has only one. The colour and streak of galena is a dull grey while molybdenite is more shiny grey and the streak has a greenish tint.

Occurrence

Molybdenite is found in many geological environments such as high-temperature hydrothermal veins and contact metamorphic rocks.

Canada's Best Localities: Spain Mine, Griffith Township, Renfrew County, Ontario; Enterprise, Sheffield County, Lennox and Addington County, Ontario; Moly Hill Mine, Malartic, Abitibi County, Québec; Bear Lake, Litchfield Township, Pontiac County, Québec.

Other Localities: Crown Point Mine, Chelan County, Washington, USA; Kingsgate, New

South Wales, Australia; Wolfram Camp, Queensland, Australia; Tae Hwa Mine, South Korea; Hirase Mine, Gifu Prefecture, Japan.

Interesting Facts

The mineral name is derived from the Greek, *molybdos*, meaning lead-like. The element molybdenum is the major constituent in molybdenite.

Molybdenite is used in lubricants. It is the main ore mineral for metallic molybdenum, which is used primarily in heat- and corrosion-resistant steel alloys. Molybdenum is also used as yellow to red pigments in paints, inks, plastics, and rubber compounds. A small amount of molybdenum is present in soils, and is important for plant growth. Plants require the element for nitrogen fixing and some animals require it to break down their food so that the body may make use of it.

Molybdenite (55595): Hexagonal, platy crystal on quartz.
Moly Hill Mine, La Motte, La Motte Tp., Abitibi Co., Québec. Width of field of view: 5 cm

STIBNITE: Antimony sulphide: Sb_2S_3

This attractive mineral forms long, slender, silvery crystals. It is the main ore of antimony. Stunning crystal groups are found in several parts of the world and they command enormous prices among collectors but they must be handled with care as they are very fragile.

Appearance

Colour: Silvery grey, often with a blackish tarnish.
Streak: Lead-grey.
Lustre and transparency: Shiny, metallic lustre and opaque.
Habit: Long slender crystals with striations along the length.

Physical Properties

Hardness: Soft (Mohs 2).
Density: Medium (4.6 g/cm^3).
Breakage: Brittle with one good cleavage face parallel to the length of the crystals.
Test: Colour, streak, and lustre are important properties.

Similar Minerals

Galena (p. 56) is more silvery than grey and has three perfect cleavages while stibnite has only one good cleavage. Galena is heavier than stibnite.
Arsenopyrite (p. 54) is harder and has striations across the diamond-shaped face.
Graphite (p. 30) is blacker with a black streak while stibnite is more grey in colour and has a grey streak. Graphite has a greasy feel.

Occurrence

Stibnite is found in low-temperature hydrothermal deposits.
Canada's Best Localities: Lake George, York County, New Brunswick; Lac Nicolet, Wolfe County, Québec; Gray Rock Mine, Lillooet District, British Columbia.
Other Localities: Manhatten, Nye County, Nevada, USA; Blackbird, Lemhi County, Nevada, USA; Baia Sprie, Romania; Ichinokawa, Shikoku Island, Japan; Hunan, China; Oruro, Bolivia.

Interesting Facts

The name stibnite refers to its antimony content. The Latin, *stibium,* is the old name for the element antimony and Sb remains its chemical symbol.

Ancient Egyptians powdered stibnite and painted it on their eyelids to make their eyes appear larger and give them a metallic, silvery grey appearance. Egyptians referred to stibnite as *stm* or "powdered antimony."

Antimony finds uses in alloys such as pewter, metal used to make printer typeface, and anti-friction metal in brakes. Its compounds are used in explosives, fireworks, and matches, in vulcanizing rubber, and in medicine to induce vomiting.

Stibnite (82157): Spray of needle-like crystals. Baia-Sprie, Maramures, Romania. Width of specimen: 5 cm

COBALTITE: Cobalt arsenic sulphide: CoAsS

Cobaltite is an important ore of the element cobalt, which has many uses, some beneficial and some not. There are several cobalt minerals, all of which are silver in colour, hard, and heavy; cobaltite is by far the most common. Cobalt minerals often have chemical alterations that give them a pinkish colour. Mineralogists refer to this colouration as "cobalt bloom."

Appearance

Colour: Silver-grey, becoming slightly pinkish with weathering.
Streak: Greyish black.
Lustre and transparency: Metallic lustre and opaque.
Habit: Most commonly massive but often as crystals in a variety of shapes: cubes, octahedra, or pyritohedra.

Physical Properties

Hardness: Hard (Mohs 5½).
Density: Heavy (6.0 g/cm^3).
Breakage: Brittle with a distinct cleavage in three directions, which gives a cube.
Test: Colour and hardness are important.

Similar Minerals

Pyrite (p. 46) has similar crystal shapes but it is brassy yellow while cobaltite is silver-grey.
Arsenopyrite (p. 54) has a darker streak and a different diamond-shaped crystal form. There is no cobalt bloom associated with arsenopyrite.

Occurrence

Cobaltite is found in high-temperature veins associated with silver, chalcopyrite, and pyrite.

Canada's Best Localities: Merry Widow Mine, Vancouver Island, British Columbia; French Mine, Similkameen District, British Columbia; Bonnet Plume River, Yukon Territory; Cobalt, Coleman Township, Timiskaming District, Ontario; Foster Township, Sudbury District, Ontario.
Other Localities: Håkansboda, Västmanland, Sweden; Tunaberg, Södermanland, Sweden; Skutterud Mines, Modum, Norway.

Cobaltite (OJ 1566): Pentagonal dodecahedron crystal. A 12-sided crystal with each face the shape of a pentagon. Håkansboda, Sweden. Width of crystal: 4 cm

Interesting Facts

The name cobaltite is from the German, *kobold*, meaning underground spirit or goblin. Cobalt was considered bewitched because early attempts at smelting it gave German smelters problems. It would not melt easily.

Cobalt-60 is an artificially produced isotope of cobalt that emits gamma rays used in cancer treatment and food sterilization. The first cobalt-60 therapy machine was built in Canada. This isotope can also be incorporated into "dirty bombs" to increase their quantity of destructive radioactive fallout.

Cobalt chloride salts are used to colour glass a beautiful deep blue.

The importance of vitamin B12 in the production of red blood cells is well known, but did you know that cobalt is an essential element in this vitamin? In fact, vitamin B12 is also called cobalamin.

Contrast the uses in cancer treatment and vitamins with that of "dirty bombs" and gunnery metal.

Cobaltite (46245): Cobaltite crystals with cube and octahedron forms. Note pink "cobalt bloom." Pnerd Claims, Bonnet Plume River, Yukon Territory. Width of specimen: 5 cm

Cobaltite (33007): Freshly broken, massive, silver-grey with a metallic lustre.
Cobalt, Coleman Tp., Timiskaming District, Ontario.
Length of specimen: 15 cm

HEMATITE: Iron oxide: Fe_2O_3

Hematite is the most common ore of iron metal. Because of its attractive red colour, it has been used in cosmetics and pigments from ancient times to the present. Tabular hexagonal crystals of hematite are termed iron roses; botryoidal forms are termed kidney ore; shiny metallic black crystals or masses are termed specularite. These varieties of hematite are sought as collector's items.

Appearance

Colour: Steel grey (crystals and specularite variety) to bright red.
Streak: Red to red-brown.
Lustre and transparency: Dull earthy to shiny, metallic (specularite variety) lustre and opaque.
Habit: Usually massive but occasionally found as pyramid-shaped crystals. Hematite can occur as rounded (botryoidal) shapes, sometimes with a radiating texture on the broken surface.

Physical Properties

Hardness: Earthy compact varieties are soft while crystals are hard (Mohs 5 – 6).
Density: Heavy (5.3 g/cm^3).
Breakage: Crystals are very brittle with no cleavage faces.
Test: Red streak, non-magnetic except as a powder.

Similar Minerals

Magnetite (p. 74) has a black streak and is highly magnetic.
Rutile (p. 118) has a more yellow to pale brown streak and is not magnetic at all.
Ilmenite (p. 78) has a more brown to black streak with less red.

Occurrence

Hematite is found as huge sedimentary deposits around the Lake Superior area, in northern Western Australia, and Itabira, Brazil. In some cases the sedimentary deposit is metamorphosed. Hematite is widespread in igneous rocks but not in large quantities.

Canada's Best Localities: Sharp 3 cm crystals from Hadley Bay, Victoria Island, Nunavut; Freilighsburg, Missisquoi County, Québec; Gooderham, Cavendish Township, Peterborough County, Ontario; Wallberg Iron Mine, Madoc Township, Hastings County, Ontario.

Other Localities: Dome Rock Mountains, La Paz County, Arizona, USA; Iron County, Wisconsin, USA; Fowler Township, St. Lawrence County, New York, USA; St. Gotthard, Switzerland; Livorno, Tuscany, Italy; Etna, Sicily, Italy; Cape Province, South Africa; Itabira, Minas Gerais, Brazil. Kidney ore from Cumbria, England.

Interesting Facts

The name hematite has its roots in the Greek word *aima*, blood, referring to its colour.

Ancient Egyptians (3000 BCE) valued iron more than silver or gold. "Native iron" came from meteorites as no method for extracting it from minerals such as hematite had yet been dis-

Hematite (34126): Specularite variety with a metallic lustre. St. John Co., New Brunswick. Width of specimen: 9 cm

Hematite (53670): Earthy, red variety with pisolitic habit. Knob Lake Junction, Labrador, Newfoundland. Length of specimen: 14 cm

Hematite (51113): Botryoidal with a shiny, sub-metallic lustre. Cumbria Co., England. Width of specimen: 9 cm

covered. Metal extraction from ores dates back to 2000 BCE in the regions of Anatolia and Persia.

When hematite (ferric oxide) is combined with other metal oxides such as nickel and cobalt they produce a set of ceramic materials called ferrites that are used extensively in computers.

The black, specularite variety of hematite is polished as decorative items and jewellery. It is sometimes called Alaska Black Diamond.

Because of its colour as a powder, hematite makes an excellent pigment in such items as "rouge" makeup.

Hematite (56453): Exceptional crystals with platy habit and deep striations forming a triangle. Cavradi Peak, Graübunden, Switzerland. Width of field of view: 6 cm

GOETHITE: Iron oxide hydroxide: FeO(OH)

Goethite is a very common and important ore of iron. It is one of the main iron ore minerals along with hematite and "limonite" in vast sedimentary deposits, like those in the Hammersly Range, Western Australia, and Lake Superior Region, Michigan, USA.

Appearance

Goethite: Massive, earthy goethite. Vermillion Cliff, Princeton, Yale District, British Columbia. Width of field of view: 10 cm

Colour: Yellowish or reddish, brown to very dark brown to almost black.

Streak: Yellowish brown.

Lustre and transparency: Dull or earthy lustre is most common but it is sometimes almost metallic (sub-metallic). Fibrous varieties are silky. It is translucent in thin fragments but mostly opaque.

Habit: Often massive but commonly in rounded, botryoidal forms that have a fibrous internal structure. May occur as small pea-shaped (pisolitic) forms and rarely as prismatic, striated crystals.

Physical Properties

Hardness: Moderate (Mohs 5).

Density: Medium (approximately 4.0 g/cm^3).

Breakage: Crystalline varieties are brittle with one perfect and one poorer cleavage and a conchoidal fracture on the non-cleavage planes.

Test: Colour, hardness, and geological occurrence are important.

Similar Minerals

"Limonite" could be considered a variety of goethite that is not crystalline (amorphous) and it occurs as an alteration (hydrated) replacement of iron minerals. "Limonite" is decidedly more yellow, fine-grained, and earthy than goethite.

Hematite (p. 68) has a red streak while that of goethite is yellow-brown.

Occurrence

Goethite is found in most parts of the earth. It forms as an alteration product of other iron minerals by oxidation or "rust." It also forms as a direct precipitation in lakes and bogs. Deep weathering of rocks in equatorial regions forms laterites rich in iron goethite or "limonite."

Canada's Best Localities: George W. McLeod Mine, Wawa area, Chabanel Township, Algoma District, Ontario; Londonderry area, Colchester County, Nova Scotia.

Other Localities: Pikes Peak, Colorado, USA; Litchfield County, Connecticut, USA; Juab, County, Utah, USA; Cornwall, England.

Interesting Facts

Goethite was named in 1789 after Johann Wolfgang von Goethe, in honour of his contributions as a scientist and poet.

Recently goethite was reported to occur at Gusev Crater on Mars. This establishes the presence, or former presence, of water on the planet. The smelting of goethite into iron is reasonably environmental friendly as the only by-products are oxygen and water.

Goethite (36907): Botryoidal goethite on white calcite with bright, brassy crystals of pyrite. Note the radiating, fibrous habit in the broken botryoidal goethite. George W. MacLeod Mine, Wawa, Chabanel Tp., Algoma District, Ontario. Width of field of view: 7 cm

Goethite (50917): Montreal Mine, Montreal, Iron Co., Wisconsin. Width of field of view: 5 cm

MAGNETITE: Iron oxide: Fe_3O_4

Magnetite is an important ore of iron. The more common iron ore, hematite (p. 68), contains less iron per unit volume than magnetite. Magnetite is the only mineral that is highly magnetic and may be a magnet itself. It is a member of the spinel group (p. 110) of minerals.

Appearance

Colour: Black.

Streak: Black.

Lustre and transparency: Dull to shiny, metallic lustre and opaque.

Habit: Often massive but crystals are common. The most common crystal form is the eight-sided octahedron.

Physical Properties

Hardness: Hard (Mohs 6).

Density: Heavy (5.2 g/cm^3).

Breakage: Brittle with conchoidal fracture, sometimes showing parting planes.

Test: It is strongly magnetic. Some samples of magnetite are magnets (lodestone variety), attracting iron filings and deflecting a compass needle.

Similar Minerals

Hematite, specularite variety (p. 68), is harder, more brittle, not as magnetic and has a red streak. Rutile (p. 118) and titanite (p. 176) are non-magnetic.

Occurrence

Magnetite is a common oxide mineral found in igneous and metamorphic rocks. It can also be found with hematite in the large sedimentary deposits of Pilbara region in Western Australia.

Canada's Best Localities: Huge masses of magnetite came from the Marmoraton open pit mine, near Marmora, Ontario. Large octahedral crystals to 15 cm from Princess Sodalite Mine, Dungannon Township, and Madawaska Mines, Faraday Township, Hastings County, Ontario; similar large crystals from Gatineau Park, Hull Township, Gatineau County, Québec; Oka, Deux Montagnes County, Québec.

Other Localities: Magnet Cove, Hot Springs County, Arkansas, USA; Balmat, Fowler Township, St. Lawrence County, USA; Franklin, Sussex County, USA; Iron Springs district, Iron County, Utah, USA; Itabira, Minas Gerais, Brazil; Gardinar Complex, Kangerlussuaq, Greenland; Binntal, Switzerland; Zillertal, Austria; Traversella, Piemonte, Italy; Kashkamar, Chelyabinskaya Oblast, Russia.

Interesting Facts

In the 16th century mariners would suspend a piece of lodestone (a variety of magnetite) from a string. One end of the piece would consistently point north. The name "lodestone"

Magnetite (50285): Rough, octahedral crystals. Madawaska Mine Property, Siddon Lake, Faraday Tp., Hastings Co., Ontario. Width of specimen: 11 cm

may have derived from an Old Norse term, *lodestar*, which was the pole star that "leads the way" in navigation.

The largest igneous magnetite deposit is in the Kiruna district, Sweden. The deposits in Marmora, Ontario, and Tasu, Moresby Island, British Columbia, are metamorphic skarn deposits. The ore at Marmora was mined as early as the 1820s and was smelted nearby. The main lode was discovered by magnetic measurements made from an airplane (airborne magnetometer) in 1955.

In 2005 an enormous magnetite-bearing sand was found in Peru.

Although most magnetite is smelted for iron, the mineral itself also finds uses. It can be used as a dense aggregate in building foundations to stabilize and reduce vibrations; also as a counterweight in excavators, ship loaders and washing machines; as a heat storage medium; and in water treatment, to remove impurities of fine particles and algae.

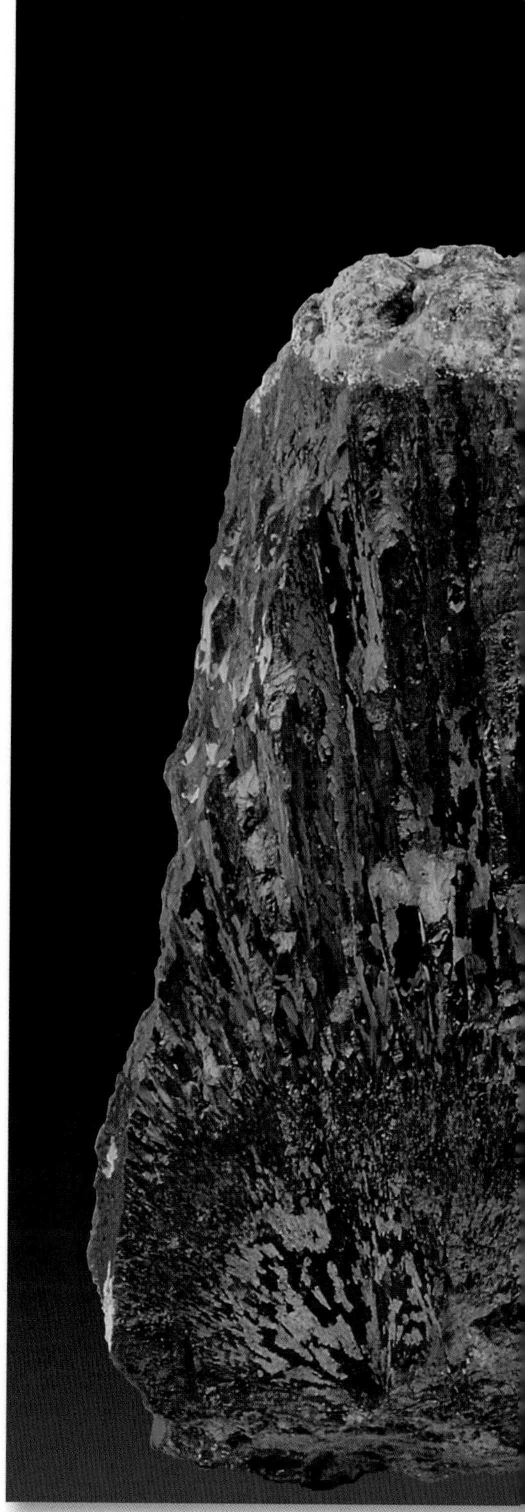

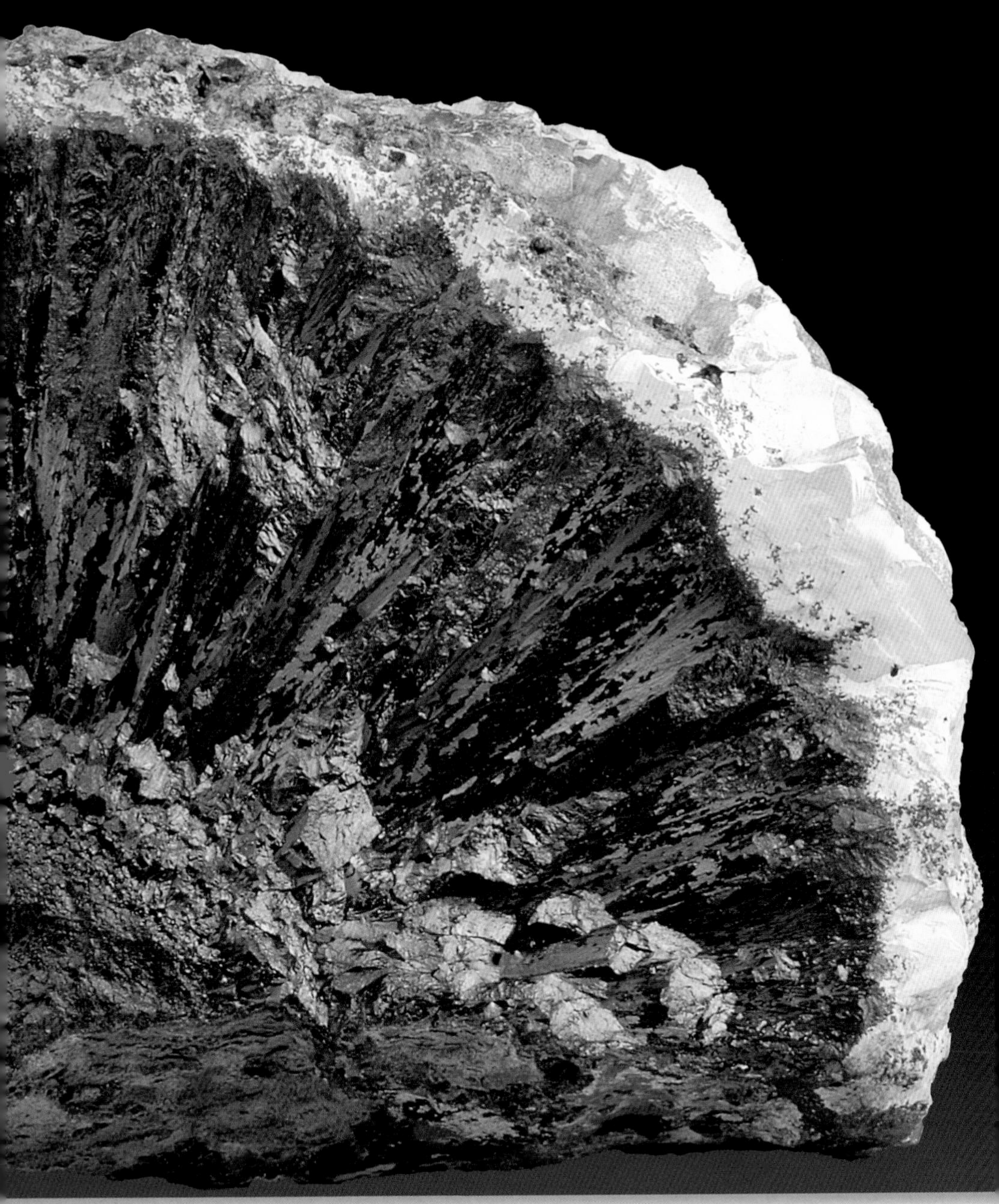

Magnetite (45587): Broken nodule with radiating structure.
J.F. Owen Prospect, Whitefish Falls, Mongowin Tp., Sudbury District, Ontario.
Width of field of view: 13.5 cm

ILMENITE: Iron titanium oxide: $FeTiO_3$

Ilmenite is one of many black, heavy minerals and care is needed to identify it. It is the most important ore of titanium, a metal of many uses. Much of the production of ilmenite comes from sand deposits in South Africa and Australia. Ilmenite weathers out of rocks and concentrates in sands because it is hard, heavy, and resistant to weathering. Since the discovery of the element titanium late in the 18th century, its importance has risen considerably, making ilmenite a valuable mineral.

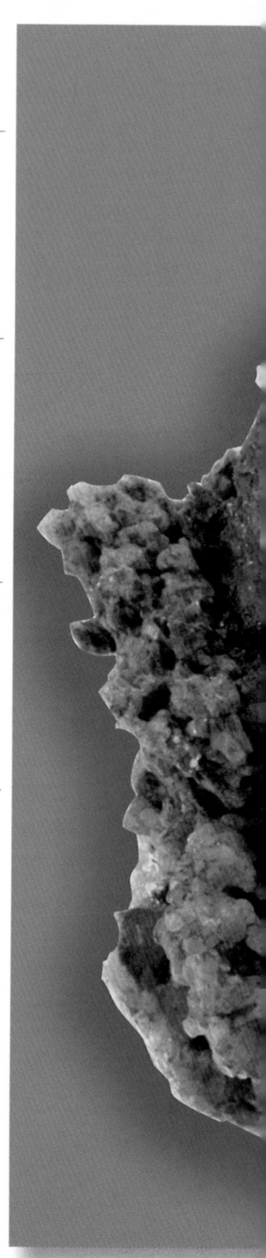

Appearance

Colour: Black or brownish black.
Streak: Black to brownish red.
Lustre and transparency: Metallic to dull lustre and opaque.
Habit: Often massive; crystals are usually tabular rhombohedra.

Physical Properties

Hardness: Hard (Mohs 5 – 6).
Density: Moderately heavy for an opaque mineral (4.8 g/cm^3).
Breakage: No distinct cleavage and an uneven to conchoidal fracture.
Test: Sometimes slightly magnetic. Hardness and streak are significant.

Similar Minerals

Hematite (p. 68) has a distinctly reddish streak.
Magnetite (p. 74) is more magnetic.
Rutile (p. 118) has a paler streak that tends to be yellowish.
Titanite (p. 176) has a white streak and crystals are wedge-shaped.

Occurrence

Ilmenite is found in igneous rocks such as diorite and gabbro, often associated with magnetite. As a placer mineral it is found in eroded sands and gravels.

Canada's Best Localities: Crystals up to 12 cm have been found at the Madawaska Mine, Faraday Township, Hastings County, Ontario; Lac Saint Jean, Lac Saint-Jean-Est County, Québec.

Other Localities: Ilmen Mountains, Russia; Arendal, Aust-Agden, Norway.

Interesting Facts

The mineral is named for the type locality (i.e., the original specimens used in the first description of this mineral were from this locality), Ilmen Lake, Ilmenski Mountains, Russia.

Titanium is an attractive, clove brown – coloured metal. The element name is for the Titans of Greek mythology who were gods of giant stature. Being lightweight, strong, and resistant to corrosion, titanium finds uses in aircraft, rockets, joint-replacement parts, and jewellery. Most of the titanium produced is used as titanium oxide in paints. It gives paint a good lustre and makes it opaque. Previously, lead had this opacity function but because of health concerns, the use of lead in paint was discontinued.

Ilmenite (34227): Tabular crystal with metallic lustre. Madawaska Mine, Siddon Lake, Faraday Tp., Hastings Co., Ontario. Width of specimen: 6 cm

CHROMITE: Iron chromium oxide: $FeCr_2O_3$

Chromite is a member of the spinel group (p. 110). Chromite is the main ore of chromium, an important industrial metal known to many of us as the bright tin-white plating on steel, chrome-plate. Chromite is not a spectacular mineral as it does not form good crystals and it is a very dull brown with a low degree of metallic lustre.

Appearance

Colour: Black.

Streak: Brown.

Lustre and transparency: Metallic lustre and opaque.

Habit: Most often as disseminated grains or as lenses, bands, or veins.

Physical Properties

Hardness: Hard (Mohs 5).

Density: Medium (4.6 g/cm^3).

Breakage: Brittle with no cleavage and a conchoidal fracture.

Test: Streak and hardness are important physical factors. Chromite is weakly magnetic; use fine grains for testing. The geological occurrence is an important association to note.

Similar Minerals

Magnetite (p. 74) has a black streak and it is much more magnetic.

Cassiterite (p. 120) has a white or near-to-white streak; it is heavier and associated with granite, not peridotite as chromite is. Often cassiterite forms crystals while chromite doesn't.

Occurrence

Chromite crystallizes early in dark (mafic) igneous rocks such as peridotite. In these rocks chromite is associated with olivine, magnetite, and corundum. As it is very resistant to alteration by high temperatures and pressures, chromite survives metamorphic conditions whereas associated minerals like olivine alter to serpentine forming serpentinite (p. 298).

Canada's Best Localities: Lac de la Blache, Saguenay County, Québec; Lac Noir, Mégantic County, Québec.

Other Localities: Lancaster County, Pennsylvania, USA; Bushveld Complex, South Africa; Ergani Maden, Turkey; Sverdlovsk region, Russia; Tiébaghi, New Caledonia; and Selukwe, Zimbabwe have notable chromite deposits.

Interesting Facts

The mineral name derives from its chemical content of chromium.

High quality stainless steel contains a significant proportion of chromium (up to 30 percent), which both hardens the steel and keeps it very bright and shiny as chromium prohibits corrosion or staining.

Just as chromite resists alteration by temperature or pressure in nature, so does the element chromium resist these physical factors. Thus it finds use in ceramics developed for lining blast furnaces.

Chromium oxide (Cr_2O_3) is a vivid green colour and is used as a pigment in paint, glass dyes, cosmetics, and military camouflage.

Chromite crystals (59740): Tiébaghi, Tiébaghi Mountains, New Caledonia. Width of field of view: 7 cm Photo: R.A. Gault

Chromite blebs in olivine (58027): Johnson Mine, Moa Bay, NW of Baracoa, Guantanamo, Cuba. Width of field of view: 8 cm Photo: R.A. Gault

PYROLUSITE: Manganese oxide: MnO_2

Pyrolusite is the most common manganese mineral and it is an important ore of this element. There are several manganese minerals that resemble pyrolusite and often they occur together as a mixture that has been termed "wad." Pyrolusite also forms fern-like patterns, some of which are enclosed in quartz and called "moss agate."

Appearance

Colour: Steel grey to black, often with a bluish tinge.

Streak: Black.

Lustre and transparency: Metallic lustre for crystals but dull for massive, granular material. Opaque.

Habit: Often massive but sometimes fibrous or columnar. Pyrolusite may grow as dendrites or branch-like forms.

Physical Properties

Hardness: Quite variable depending on the form. Crystals are quite hard (Mohs 6) but the more common massive varieties can be very soft (Mohs 1 – 2) and readily marks paper or hands like graphite.

Density: Medium (about 5.0 g/cm^3).

Breakage: Crystals are brittle with a distinct cleavage. Massive forms are soft and sooty.

Test: Sooty nature of massive forms is important. It leaves black marks on your hands and feels dry.

Similar Minerals

Graphite (p. 30) feels greasy while pyrolusite feels dry and sooty. Crystals of pyrolusite are much harder than graphite.

Occurrence

Pyrolusite is found in highly oxidizing conditions. It is often the product of weathered manganese minerals. Often several manganese oxide minerals are intimately intermixed and this is called "wad." Wad is actually quite common as the major constituent of nodules found on the seafloor. Nodules of only a centimetre in size take several million years to grow.

Canada's Best Localities: Steep Rock Mines, Freeborn Township, Rainy River District, Ontario; Markhamville, Sussex Parish, Kings County, New Brunswick; Bridgeville, East River, Pictou County, Nova Scotia.

Other Localities: Casa Grande, Pinal County, Arizona, USA; Braga County, Michigan, USA; Mapimi, Durango, New Mexico, USA; Broken Hill, New South Wales, Australia; Tarn, France; Horni Blatna, Czech Republic; Rosbach, Westphalia, Germany; Kisenge, Katanga, Democratic Republic of Congo.

Pyrolusite
Thüringen, Germany.
Width of specimen: 5 cm

Interesting Facts

The mineral name, pyrolusite, is from Greek, *pyr* meaning fire and *louxo,* meaning to wash. As early as 17,000 years ago manganese dioxide was in use in pigments. Umber, a mixture of limonite and pyrolusite, was extensively used in the paintings by Vermeer and Rubens.

Egyptians and Romans discovered that this mineral used as an additive washed out tints of green and brown in molten glass. Excessive amounts of pyrolusite tint glass an amethyst colour.

Large nodules of pyrolusite or "wad" have been dredged up from the ocean floor. In the future this may become an important source for manganese metal but at present it is too expensive to mine this source. Manganese is an important additive in making steel alloys, as it hardens the steel without making it more brittle.

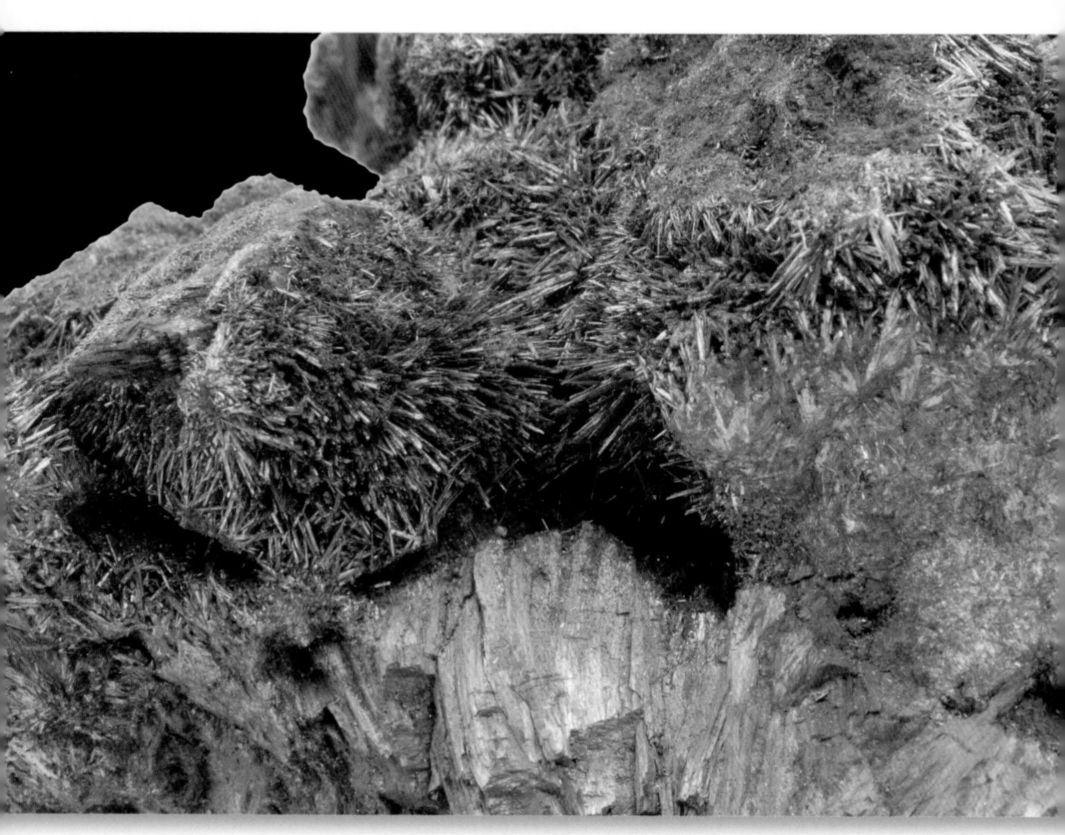

Pyrolusite: Wadberget, Västergötland, Sweden.
Width of field of view: 5 cm

Non-metallic Minerals

DIAMOND: Carbon: C

Diamond is a mineral that you will probably only find in jewellery shops. By far the most popular gemstone, it is also the hardest naturally occurring material and as such, has many industrial uses. Although diamond is extremely hard it is also very brittle, breaking rather easily along cleavage planes.

Appearance

Colour: Colourless, yellow, brown, and rarely blue, pink, and green. It can sometimes be grey or black, called bort, due to inclusions.

Streak: Too hard to get a streak but the powder is white.

Lustre and transparency: Adamantine or greasy lustre; transparent to translucent.

Habit: Usually as crystals, often with octahedral form and sometimes as cubes, tetrahedra, or 12-sided dodecahedra.

Diamond: Octahedron with growth patterns on octahedral face. Ekati Mines, Lac de Gras, Northwest Territories. Width of crystal: 0.5 cm

Diamond (56776): Flattened, octahedral crystal on kimberlite rock. Mir pipe, Mirnyy, Sakha Republic, Russia. Length of crystal: 1.5 cm

Physical Properties

Hardness: Hard (Mohs 10). The Mohs scale is not linear; it is only a relative guide to hardness. If it were linear, diamond would be approximately 42 compared to corundum which has a Mohs hardness of 9.

Density: Moderate (3.5 g/cm^3).

Breakage: Brittle with a distinct octahedral cleavage (i.e., three good cleavages forming an octahedron). Diamond has an uneven fracture on the non-cleavage planes.

Test: Lustre and hardness.

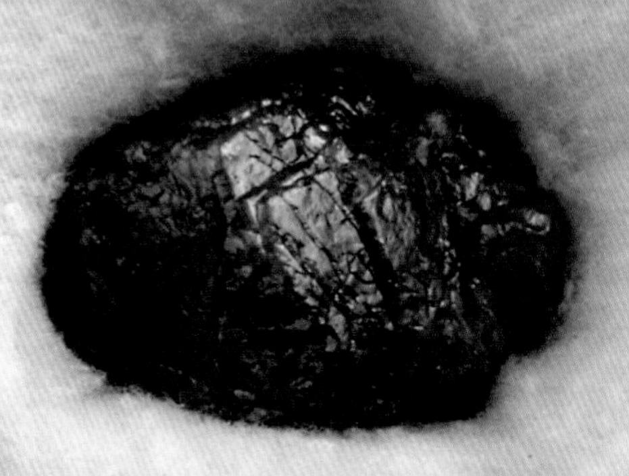

Similar Minerals

Quartz (p. 218) is softer (Mohs 6), and has no cleavage while diamond has three good cleavages.

Occurrence

Diamond is found in igneous rocks that form at great depths in the Earth's crust (150 km or deeper). The high pressure and temperature give rise to this very special form of carbon.

Canada's Best Localities: The Diavik and Ekati Mines in the Northwest Territories, 300 km north of Yellowknife, produce high-quality diamonds up to 2 carats when cut.

Other Localities: Central and southern Africa have always been important producers. There are also significant diamond deposits in Australia, Brazil, India, and Russia.

Interesting Facts

The mineral name is a corruption of the Greek, *adamas*, meaning invincible, and alluding to the hardness of the mineral. Diamond is mentioned in ancient Indian texts as early as 296 BCE. At that time there were alluvial deposits of diamond in various parts of India.

Graphite and diamond are two structural forms (dimorphs) of carbon that occur in nature. The extreme differences between these dimorphs can be explained by the nature of the atomic bonds in the minerals.

Some diamonds are harder than others, varying by as much as 40 percent in absolute hardness. This variance is due to differing growth histories. A crystal that is grown in one stage has fewer inclusions and flaws and hence is harder than a crystal having multiple growth stages.

Famous diamonds include: the largest gem quality rough, the Cullinan Diamond, weighed 3,107 carats. The cut Cullinan I, which is in the British crown jewels, weighs 530 carats. The largest cut diamond, the Golden Jubilee, 546 carats, is a yellow-brown diamond and the largest flawless and colourless diamond is the Millennium Star, at 203 carats.

Black bort or carbonado diamond with several octahedra, one of which is flattened into a trilling. It takes approximately a 1 carat crystal to cut a 0.5 carat stone.
Ekati Mines, Lac de Gras, Northwest Territories.
Width of field of view: 6 cm

SULPHUR: S

Sulphur has been recognized since antiquity due to its distinctive appearance. Albertus Magnus, a 13th-century scholar, recognized "live" sulphur and "fused" sulphur. "Live" sulphur is the naturally occurring sulphur described here, while "fused" sulphur is that left after smelting metal from sulphide minerals.

Appearance

Colour: Yellow, sometimes having a green or brown tinge.
Streak: White.
Lustre and transparency: Greasy to resinous.
Habit: Usually massive but often occurs as rhombic, diamond-shaped crystals.

Physical Properties

Hardness: Soft (Mohs 2).
Density: Light (2.1 g/cm^3).
Breakage: Brittle with an uneven or conchoidal fracture.
Test: Colour and low hardness are usually sufficient to distinguish this mineral. It melts at low temperature and has a distinct smell when crushed or heated.

Sulphur: Rhombic dypyramid with basal pinacoid crystals on quartz. Sicily, Italy. Width of field of view: 15 cm

Similar Minerals

Fluorite (p. 106), baryte (p. 132), topaz (p. 172), and feldspar (p. 226) may be yellow, but all of these minerals are considerably harder than sulphur.

Occurrence

Sulphur is found in regions of recent volcanic activity. In sedimentary gypsum and limestone formations, sulphur may be found as a result of bacterial separation of sulphates. Sulphur is also found in the salt domes above oil and gas reservoirs.

Canada's Best Locality: No significant specimens.

Other Localities: Baja California, Mexico. Beautiful crystals come from Argento, Sicily, and Perticura, Romagna District, Italy.

Interesting Facts

The mineral name sulphur is derived from the Latin, *sulphurum*, probably from a root meaning to burn. Usage of the word dates back to 2000 BCE and replaces the term brimstone or burning stone.

Millions of tonnes of sulphur are produced annually for the chemical industry. Applications include sulphuric acid, pesticides, and the vulcanizing of rubber. Sulphur is produced primarily as a by-product of refining oil and gas. It also comes from the smelting of sulphide ores such as pentlandite, sphalerite, and galena.

SPHALERITE: Zinc sulphide: ZnS

The mineral name is from the Greek, *sphaleros*, meaning mistaken or deceiving, because the mineral was often mistakenly identified. Sphalerite is difficult to identify because of large variations in colour, transparency, and lustre.

Appearance

Colour: Quite variable from pale yellow, through brown to almost black. Adding more iron in this mineral causes the colour to become darker.

Streak: Yellow to brown.

Lustre and transparency: Greasy to almost metallic lustre and transparent to opaque. Its lustre is often described as resinous because it is like the reflection of light from pine gum or resin.

Habit: Often as complex crystals but in many localities sphalerite occurs as cleavable masses or grains.

Sphalerite (32300): Rounded crystal with greasy lustre on limestone. Lincoln Quarries, Beamsville, Clinton Tp., Lincoln Co., Ontario. Width of field of view: 3 cm

Sphalerite (32310): Massive with many cleavages, sub-metallic lustre.
Wild Goose Point area, Mac Gregor Tp., Thunder Bay District, Ontario.
Width of specimen: 9 cm

Physical Properties

Hardness: Medium hardness (Mohs 3 – 4).
Density: Moderately heavy (4.0 g/cm^3).
Breakage: Brittle with distinct cleavage planes or a conchoidal fracture.
Test: Streak, lustre, and density are important properties.

Similar Minerals

Some dark-coloured amphiboles (p. 196) like hornblende or actinolite may resemble dark sphalerite but they are harder (Mohs 6), and have a glassy lustre.

Occurrence

Sphalerite is most commonly formed by warm to hot hydrothermal solutions. It is often associated with galena, chalcopyrite, and pyrite. Rarely is it found in sedimentary rocks.

Canada's Best Localities: Polaris Mine, Little Cornwallis Island, Nunavut; Nanasivik Mine, Baffin Island, Nunavut; Pine Point, Mackenzie District, Northwest Territories; Bluebell Mine, Kootenay District, British Columbia; Lucky Jim Mine, Zincton, British Columbia; Beamsville, Clinton Township, Lincoln County, Ontario; Mont Saint-Hilaire, Rouville County, Québec; Brunswick # 12 Mine, Bathurst, Gloucester County, New Brunswick.

Other Localities: Grand County and Mineral County, Colorado, USA; Baxter Springs, Kansas, USA; Hardin County, Illinois, USA; Joplin, Missouri, USA; St. Lawrence County, New York, USA; Ottawa County, Oklahoma, USA; Chester County, Pennsylvania, USA; Carthage, Tennessee, USA; Naica, Chihuahua, Mexico; Trujillo, Peru; Cornwall and Cumbria, England; Wanlockhead, Scotland; La Calamine, Belgium; Santander, Spain; Cavnic, Romania; Trepca, Serbia; Madan, Bulgaria.

Interesting Facts

Zinc was not recognized as a metal until the 16th century. Earlier attempts to smelt sphalerite were unsuccessful. Today sphalerite is the main ore of zinc. Zinc metal is used extensively to galvanize or coat iron and steel to prevent rusting. It is an essential addition or alloy in brass and other metals. The human body requires zinc as an essential constituent of enzymes that metabolize carbon dioxide and proteins. Zinc fertilizer is sometimes used to treat soils deficient in the mineral.

Sphalerite (51331): Rounded tetrahedron crystal with dusting of brassy pyrite.
Nanisivik Mine, Nanisivik, Baffin Island, Nunavut.
Length of field of view: 7 cm

CINNABAR: Mercury sulphide: HgS

The most notable property of cinnabar is its scarlet red colour. At one time vermilion was used as a synonym for this mineral. Cinnabar has been used as a pigment since ancient times and today it is often referred to as China Red, since much of the raw material comes from China.

Appearance

Colour: Red to brownish.
Streak: Red.
Lustre and transparency: Adamantine crystals to dull in earthy varieties. Transparent to translucent in splinters.
Habit: Often fine-grained or massive, rare crystals are rhombic or diamond-shaped.

Physical Properties

Hardness: Fairly soft (Mohs 2).
Density: Heavy (8.1 g/cm^3).
Breakage: Brittle with a distinct cleavage plane.
Test: Colour and hardness are distinctive.

Similar Minerals

Hematite (p. 68) is harder (Mohs 5), and is slightly magnetic.
Cuprite (p. 116) is harder (Mohs 3½) and more of a black-red colour.

Occurrence

Cinnabar is found in low-temperature veins near volcanic rocks or hot spring areas. Often it is associated with native, liquid mercury.

Canada's Best Locality: Pinchi Lake Mine, British Columbia.
Other localities: Kuskokwin River, Alaska, USA; Lovelock, Pershing County, Nevada, USA; Moschellandsberg, Germany; Almadén, Spain. Excellent crystals from Hunan Province, China.

Cinnabar (0J1506): Twinned crystal on quartz.China. Width of field of view: 2 cm

Interesting Facts

The name cinnabar is possibly from the Arabic *zibjafr* or Persian *zinjifrah*, meaning dragon's blood and referring to the colour of the mineral. The chemical symbol for mercury, Hg, is from a Latinized Greek word, *hydrargyprum*, meaning watery or liquid silver.

Cinnabar is the main ore of mercury. Mercury is one of four metals that are liquid at normal temperatures. It is used in thermometers and electrical switches. An electrical discharge through mercury produces a bluish glow in the shorter wavelengths (less than 280 nanometres) of the ultraviolet spectrum. This property makes mercury useful in the manufacture of ultraviolet lights.

Mercury is still used in dental fillings as an amalgam with an alloy of silver, copper, and tin but this use is controversial, as mercury is a neurotoxin.

Cinnabar (32288): Massive habit with a dull, earthy lustre. British Columbia. Length of specimen: 6 cm Photo: R.A. Gault

Cinnabar (82250): Massive habit with an adamantine lustre. Monte Amiata, Tuscany, Italy. Width of field of view: 9 cm

HALITE: Sodium chloride: NaCl

Halite, the mineral we know as common salt, is an essential part of our diet. Egyptian art work from as early as 1450 BCE records salt making. Salt is important in every civilizations' history. Ancient Greece exchanged salt for slaves, giving rise to the expression, "not worth his salt." Roman soldiers received special salt rations known as "salarium argentum" which gave rise to the English word "salary."

Appearance

Colour: Colourless to white, sometimes with tints of grey, yellow, orange, or blue.
Streak: White.
Lustre and transparency: Vitreous or glassy lustre and transparent to translucent.
Habit: Cubic crystals, sometimes with stepped faces or massive.

Physical Properties

Hardness: Soft (Mohs 2).
Density: Light (2.2 g/cm^3).
Breakage: Brittle with a distinct cleavage in three directions, each at 90º to each other, making it a cubic cleavage.
Test: Salty taste, dissolves in water, perfect cubic cleavage.

Similar Minerals

Sylvite (p. 104) is bitter in taste and orange in colour.
Gypsum (p. 124) has only one perfect cleavage.

Occurrence

Halite is found in areas of evaporated sea water often with beds of gypsum, anhydrite, calcite, and sometimes sylvite. In arid desert conditions it is found as salt lakes. Associated with volcanic regions, it may be found as salt springs.

Canada's Best Localities: Huge, massive deposits are mined in Esterhazy, Saskatchewan; Windsor, Ontario; Sussex, New Brunswick; and Pugwash, Nova Scotia.
Other Localities: Carlsbad, New Mexico, USA; Mentor, Ohio, USA; Cayuga, New York, USA; Salton Sea and Searles Lake, California, USA. Purple to blue crystals from Stassfurt and Hessen, Germany; Mulhouse, France. Excellent crystals from Poland.

Halite (49468): Stacked, cubic crystals.
Inowroclaw Mine, Inowroclaw, Poland.
Width of specimen: 10 cm

Interesting Facts

The mineral name is from the Greek *hals* meaning salt.

The uses of halite or salt are well known in our home but did you know that there are also many industrial applications for this mineral? Industries such as pulp and paper, textile dyes, rubber manufacturing, metal processing, ceramic manufacture, soap making, and leather tanning all use salt.

Much of the halite mined in northern climates goes onto the roads in the winter. Salt lowers the freezing point of water making it possible to reduce the amount of ice down to temperatures of –21°C, but below this temperature even salty water freezes.

Halite (84694): Massive with perfect cubic cleavage.
Alwinsal Mine, Lanigan, Saskatchewan.
Width of specimen: 8 cm

SYLVITE: Potassium chloride: KCl

Sylvite has the same crystal structure (isostructural) as halite, NaCl. Both salts are deposited from evaporating sea water. Sylvite is the more commercially important salt. Because potassium is an essential element in plant growth, sylvite is mined for the fertilizer industry.

Appearance

Colour: Colourless or white; sometimes coloured due to inclusions; often reddish due to hematite impurities.

Streak: White.

Lustre and transparency: Vitreous and transparent to translucent.

Habit: Often massive, sometimes as crystals with cubic and/or octahedral faces.

Physical Properties

Hardness: Soft (Mohs 2).

Density: Light (2.0 g/cm^3).

Breakage: Brittle, somewhat sectile. Sylvite has three cleavages at 90° to each other (perfect cubic cleavage).

Test: Sylvite tastes bitter and is sectile. It is readily soluble in water.

Similar Minerals

Halite (p. 100) tastes salty while sylvite is bitter. Sylvite is more sectile when cut with a knife, while halite powders when cut.

Occurrence

Sylvite is found in marine evaporite deposits. It crystallizes later than halite as it is more soluble in water and it is much rarer than halite. It commonly occurs with halite, gypsum, and anhydrite.

Canada's Best Localities: Huge deposits of massive sylvite beds occur in Saskatchewan and near Sussex, New Brunswick.

Other Localities: Mesa Verde, Arizona, USA; Carlsbad, New Mexico, USA; Salton Sea, California, USA; excellent crystals from Stassfurt, Germany.

Interesting Facts

Sylvite was named in honour of Françoise Sylvius de la Boe, a physician and chemist from Leyden, Netherlands. It was first described in 1832 after being discovered at Mt. Vesuvius, Italy.

Sylvite ore is commonly referred to as potash, since much of the ore is used to produce potassium hydroxide. North American Indians knew the value of potash. They boiled wood ashes, which are rich in potassium, in water, to produce potassium hydroxide, and then put the mixture on their fields to increase plant growth.

Sylvite (32209): A mass of cubic crystals. Saskatoon, Saskatchewan. Width of specimen: 9 cm

FLUORITE: Calcium fluoride: CaF_2

Fluorite is certainly one of the most beautiful minerals. It comes in a variety of colours and in perfect crystals. Crystals with complete octahedral faces are rare but fluorite has perfect octahedral cleavage and specimens are often sold as "crystals" even though the face is not a natural form but the result of cleaving. Blue John, a fluorite variety from Derbyshire, England, was widely used in ornamental vases and other decorative objects beginning around 1750.

Appearance

Colour: Colourless to brown to almost black, shades of red, pink, yellow, green, blue, and violet. Colour is commonly zoned parallel to crystal faces.

Streak: White.

Lustre and transparency: Vitreous lustre and transparent to translucent.

Habit: Massive or crystals that are cubic and/or octahedral.

Physical Properties

Hardness: Medium (Mohs 4).

Density: Medium (3.2 g/cm^3).

Breakage: Brittle with a perfect octahedral cleavage.

Test: Cleavage is distinct and it is noticeably heavy for a non-metallic mineral.

Similar Minerals

Calcite (p. 138) is softer (Mohs 3), lighter density (2.7 gm/cm^3), has perfect rhombohedral cleavage in all three planes, and fizzes in dilute acid.

Baryte (p. 132) is heavier with a density of 4.5 gm/cm^3.

Fluorite (34793): These are not crystal forms but octahedral cleavage fragments. Varying shades of purple. Rosiclare, Hardin Co., Illinois. Length of field of view: 9 cm

Occurrence

Fluorite is found in a wide variety of rock types from granite to hydrothermal veins to vugs in sedimentary limestone.

Canada's Best Localities: Purple octahedra from the Rock Candy Mine, near Grand Forks, British Columbia; pale, clear green crystals to 10 cm from Madoc, Hastings County, Ontario; purple cubes edged with yellow from Rossport, Thunder Bay District, Ontario gives a

Fluorite (35453): The translucent, almost colourless fluorite crystal has a cube crystal form, frosted, almost square faces and a large octahedral form with smooth somewhat triangular face. White baryte is the crusty, botryoidal base. Rogers Mine, Madoc, Huntingdon Tp., Hastings Co., Ontario. Length of field of view: 5 cm

subtle insight into the growth pattern of a crystal; Dundas Quarry, West Flamborough Township, Wentworth County, Ontario; very complex blue-green crystals from Old Chelsea, Hull Township, Gatineau County, Québec; Mount Pleasant Mine, Charlotte County, New Brunswick; cubic crystals to 30 cm from St. Lawrence, Newfoundland.

Other Localities: Hardin County, Illinois, USA; Allen County, Indiana, USA; Elmwood Mine, Tennessee, USA; Jefferson County, New York, USA; Chihuahua and Coahuila, Mexico; Cumbria, Durham, and Cornwall, England; Freiburg, Germany; Oviedo, Spain; Taihang-Shan Mountains, China.

Interesting Facts

The mineral name fluorite is from the Latin, *fluere*, meaning to flow. It was first used by miners in the time of Georgius Agricola (1529) and refers to melting the mineral in a furnace "like ice in the sun." Fluorite is still used today as a flux to lower the melting temperatures necessary to manufacture steel and aluminum.

Fluorite is used to make opalescent glass, enamels for cookware, and in the production of hydrofluoric acid. Because of its low light dispersion (breaking white light into its component colours) fluorite is also used in the lenses of expensive cameras and telescopes to minimize distortion of the image.

Fluorite (48648): Octahedral crystal form. The dark colour is induced by radiation from this deposit. Madawaska Mine, Bancroft, Faraday Tp., Hastings Co., Ontario. Width of specimen: 4 cm

Fluorite (34910): A crystal showing the cube face on top and triangular octahedral faces cutting the cube corners. The octahedral cleavage is evident as internal fracture planes.
Deloro, Marmora Tp., Hastings Co., Ontario.
Width of field of view: 9 cm

Fluorite (48304): A mass of clear, colourless cubes.
Amherstburg Quarry, Amherstburg, Malden Tp., Essex Co., Ontario. Width of field of view: 5 cm

SPINEL: Magnesium aluminum oxide: $MgAl_2O_4$

Spinel is the name of a mineral species and a mineral group. In this book we cover three members of the group; spinel, magnetite, and chromite. For centuries red spinel has been prized as a gem, rivalling ruby in colour but not in value. Many historically famous objects contained gems believed to be rubies, but they were in fact spinel.

Appearance

Colour: Pure spinel is red, but it can range through colourless, green, blue, brown, and black.

Streak: Powder is white.

Lustre and transparency: Vitreous lustre and transparent to translucent.

Habit: Often as crystals in perfect octahedra (eight triangular sides as two pyramids sharing a common base). It may also be massive or granular.

Spinel (80785): Gemmy octahedral crystals. Upper Mogok River, Katha District, Sagaing, Myanmar. Length of field of view: 2.5 cm

Physical Properties

Hardness: Hard (Mohs 7 – 8).

Density: Medium (3.6 g/cm^3).

Breakage: Difficult to break and it has no cleavage and an uneven fracture.

Test: Crystal form and hardness are important.

Similar Minerals

Corundum (p. 112), ruby variety, is hexagonal in outline while spinel is square. Ruby is red compared to spinel, which is more pinkish. Corundum may have a parting while spinel does not. Corundum is considerably harder (Mohs 9) than spinel but spinel is also one of the hardest minerals known.

Occurrence

Spinel is found in igneous rocks such as basalt, and the metamorphic equivalents of these igneous rocks, serpentinites, as well as metamorphosed limestones (marble).

Canada's Best Localities: Glencoe Island, Hudson Strait, Nunavut; Bathurst Township, Lanark County, Ontario; Parker Mine, Bigelow Township, Labelle County, Québec.

Other Localities: Franklin, New Jersey, USA; Amity, New York, USA; Mogok, Myanmar; Yakutia, Russia; Sri Lanka; large crystals to 12 cm from Betrok, Madagascar.

Interesting Facts

The mineral name spinel is derived from Latin, *spina*, meaning thorn, for its sharp octahedral crystals.

The Black Prince's Ruby is one of the oldest of the Crown Jewels of the United Kingdom. It centres the Imperial State Crown, just above the Cullinan II diamond. It has been in the possession of the British kings since it was given in 1367 to Edward of Woodstock (the "Black Prince"). However, the 170 carat (34 g) stone is not a ruby, but a spinel. It is thought to be from the Badakhshan mines along Afghanistan's border.

Spinel (49612): Octahedral crystal in calcite.
Parker Mine, Notre-Dame-du-Laus, Bigelow Tp., Labelle Co., Québec. Width of field of view: 4.5 cm

CORUNDUM: Aluminum oxide: Al_2O_3

Corundum is best known for its gem varieties – ruby and sapphire. Ruby is red corundum while sapphire is any other coloured, gemmy corundum: blue, yellow, green, purple, or pink. Star sapphire is a corundum crystal that has fine, needle-like inclusions (usually rutile) oriented during crystal growth into a six-pointed star.

Appearance

Colour: Usually grey but many colours are found and sometimes it may be a dark red-brown. The gem quality examples receive variety names ruby (red corundum) and sapphire (usually blue corundum but it can also be green, yellow, purple, white, or colourless).

Streak: White when crushed. It is too hard to get a streak on a streak plate.

Lustre and transparency: Vitreous lustre, sometimes it has a slightly greasy appearance due to alteration of the crystal's surface. Corundum is transparent to translucent.

Habit: Often as hexagonal, barrel-shaped crystals. Sometimes striation lines are seen across crystal faces parallel to the base. Also granular and massive habits occur.

Corundum (56779): Hexagonal dipyramid with striations. Sapphire variety. Balangoda, Ratnapura District, Sri Lanka. Length of crystal: 9 cm

Corundum (30052): Sapphire variety, note hexagonal shape of the rough crystal outline. Bancroft area, Dungannon Tp., Hastings Co., Ontario. Width of specimen: 5 cm

Corundum (35470): Barrel shaped, hexagonal prisms. Ruby variety. India. Width of field of view: 12 cm

Physical Properties

Hardness: Very hard (Mohs 9).
Density: Quite heavy for a non-metallic mineral (4.0 g/cm^3).
Breakage: Brittle with an uneven fracture. Crystals are often twinned and this can give a parting or stepped appearance on the base of the crystal.
Test: Density and hardness are important properties.

Similar Minerals

Quartz (p. 218) is softer (Mohs 7) and lighter (2.65 g/cm^3). It does not have the basal parting or greasy appearance of corundum.

Occurrence

It is found in silica-poor igneous rocks such as syenite or in metamorphic rocks. Its hardness resists weathering, resulting in corundum sand and gravel deposits.

Canada's Best Localities: York River, Dungannon Township, Hastings County, Ontario; Craigmont Corundum Mine, Raglan Township, Renfrew County, Ontario; Rosenthal, Brudenell Township, Renfrew County, Ontario.
Other Localities: The finest rubies come from Myanmar (formerly Burma) and India is noted for its sapphires. Important deposits of gem corundum include: Ratnapura, Sri Lanka; Ruby Harts Range, Northern Territory, Australia; Ihosy, Madagascar; Arusha, Tanzania; Mysore, India; Transvaal, South Africa; Chiridzi, Zimbabwe.

Among the notable rubies and sapphires: The Edward's Ruby (167 carats) is displayed at the Britian's Natural History Museum. The Star of India, a star sapphire (563 carats) from Sri Lanka, and the Delong Star Ruby (100 carats) from Burma are displayed at the American Museum of Natural History.

Interesting Facts

The mineral name is derived from Sanskrit, *kuruvinda*, meaning ruby. Due to corundum's hardness it is important as an abrasive in machining large equipment parts and as the grit in sandpaper. Emery paper is an impure variety of corundum.

Corundum (33624): Hexagonal prisms in nepheline.
Rosenthal, Brudenell Tp., Renfrew Co., Ontario.
Width of field of view: 11 cm

Corundum (A2005-027):
Ambiandono, Ambositra
Region, Fianarantsoa
Province, Madagascar.
Width of field of view: 6 cm

CUPRITE: Copper oxide: Cu_2O

Cuprite is a minor ore of copper. It is an important indication of the geology as it occurs in a near surface, oxidized deposit whereas the element copper is usually in sulphide minerals.

Cuprite is highly prized as a mineral specimen when it displays a deep red colour and fine crystals.

Appearance

Colour: Dark red to almost black.

Streak: Brownish red.

Lustre and transparency: Submetallic or adamantine lustre, sometimes earthy. It is translucent but almost opaque.

Habit: Often as crystals: octahedral and cube forms are common. It may also be massive or as fuzzy crusts.

Physical Properties

Hardness: Moderate (Mohs 3½ – 4).

Density: Heavy (6.1 g/cm^3).

Breakage: Brittle with a distinct but rather poor cleavage in four directions (octahedral cleavage) and an uneven fracture on the non-cleavage planes.

Test: Colour, streak, and softness are important.

Similar Minerals

Hematite (p. 68) is harder (Mohs 6) than cuprite and weakly magnetic which cuprite is not. Cinnabar (p. 96) is softer (Mohs 2½) than cuprite and the red colour is more pronounced.

Occurrence

It is found in near surface, or oxidized, portions of copper deposits associated with copper, malachite, azurite, and limonite.

Canada's Best Localities: Britannia Mine, New Westminister District and Valley Copper Mine, Highland Valley, Kamloops District, British Columbia.

Other Localities: Bisbee, Cochise County, Arizona, USA; Santa Rita, New Mexico, USA; Wheal Gorland, Cornwall, England; Chessy, Rhone Department, France; Tsumeb Mine, Namibia; Windhoek, Namibia; Shaba, Zaire; Kolwezi, Katanga, Democratic Republic of Congo.

Interesting Facts

The mineral name is from Latin, *cuprum*, in reference to its copper content. The word *cuprum* originated from *aes cyprium*, or metal from Cyprus, where copper was mined to supply the Roman Empire.

Many cuprite specimens are coated with green malachite. You should not remove this coating as it compromises the true nature of the specimen.

The uses of copper are given with the native copper description (p. 20).

Cuprite (30058): Crystals with adamantine lustre and translucent.
Tsumeb Mine, Tsumeb, Otjikoto, Namibia.
Length of field of view: 9 cm

RUTILE: Titanium oxide: TiO_2

Rutile is relatively rare but important as a source of titanium dioxide and the metal titanium. It is the most common of three titanium dioxide polymorphs, the other two being brookite and anatase.

Rutile is visible as an inclusion in quartz, but invisible except for the optical effect of an apparent star in corundum. In star ruby and star sapphire (gem varieties of corundum) very fine, oriented needles of rutile in the crystals reflect light in a polished gem, giving star-like optical properties known as asterism.

Appearance

Colour: Golden brown, reddish brown to black.

Streak: Yellowish or very pale brown.

Lustre and transparency: Adamantine or sub-metallic lustre and transparent in thin fragments to translucent.

Habit: Long, prismatic crystals terminated by a pyramid are often striated parallel to the prism face. Crystals may be twinned and rutile is sometimes an inclusion in rutilated quartz. Often occurs as fine needles the diameter of a hair, and sometimes as coarse granules.

Rutile (34338): Multiple twinned crystal with deeply striated faces. Minas Gerais, Brazil. Width of field of view: 8.5 cm

Physical Properties

Hardness: Hard (Mohs 6 – 6½).

Density: Medium (4.2 g/cm^3).

Breakage: Brittle with a distinct cleavage and an uneven fracture on the non-cleavage planes.

Test: Colour and hardness are important.

Similar Minerals

Ilmenite (p. 78) has no cleavage and crystals are tabular. Ilmenite occurs in igneous rocks while rutile is most often found in metamorphic rocks.

Magnetite (p. 74) or specular hematite (p. 68) is magnetic or slightly magnetic.

Titanite (p. 176) has a white streak and crystals are wedge-shaped, not elongate like rutile.

Zircon (p. 178) has a white streak and crystals have no striations and are square in cross-section.

Occurrence

Rutile is commonly found in metamorphic rocks such as schist or gneiss. In igneous rocks it is found in granites or granite pegmatite and syenite.

Canada's Best Locality: Little Dam Lake, Templeton Township, Hull County, Québec.

Other Localities: Magnet Cove, Hot Springs County, Arkansas, USA; Graves Mountain, Lincoln County, Georgia, USA; Parkesburg, Chester County, Pennsylvania, USA; Stoney Point, Alexander County, North Carolina, USA; Binntal, Switzerland; Novo Horizonte, Bahia, Brazil. Rutilated quartz comes from Minas Gerais, Brazil, and Madagascar.

Interesting Facts

The mineral name rutile is from the Latin *rutilas*, describing its red colour.

Rutile, or its synthetic equivalent, is non-toxic and reflects and scatters light. It is used extensively in high-quality paint pigments to replace toxic lead oxides. In paper and plastics it adds brilliance and sheen. Small particles of rutile are transparent to visible light but reflect ultraviolet rays making it useful as sunscreen. Crystals of synthetic TiO_2 are used to simulate diamonds and are sold under the name "Titania."

Titanium metal is lightweight, incredibly strong, and resistant to high temperature and corrosion. These properties find uses in aerospace materials, hip replacements, pacemakers, eyeglass frames, and electric welding rods.

Rutile (40945): Striated prismatic crystals with striations parallel to the length and good cleavage on feldspar. Little Dam Lake, Petit Lac du Barrage, Templeton Tp., Hull Co., Québec. Width of field of view: 8 cm

CASSITERITE: Tin oxide: SnO_2

Cassiterite is the main ore of tin and has been mined as such for centuries. Although it is another heavy, dark mineral, specimens are unusually attractive due to its sparkle caused by its high lustre and multi-faced crystals. Cassiterite is not a common mineral, which accounts for the high cost of tin, but it is worth knowing because of its value as an ore.

Appearance

Colour: Dark reddish brown to brown-black.

Streak: White or pale grey to pale yellow.

Lustre and transparency: Adamantine lustre on crystals or dull on weathered samples. It is translucent to opaque.

Habit: Often crystals are prismatic or tabular. Prisms are short and can be four- or eight-sided. Sometimes the prism is topped by a pyramid. Twinned crystal aggregates are common and if the twin keeps repeating it can grow in a cyclical fashion (360º rotation). It may also be massive, granular, or in rounded habits. Because of its appearance, the rounded or concretionary cassiterite is called "wood tin."

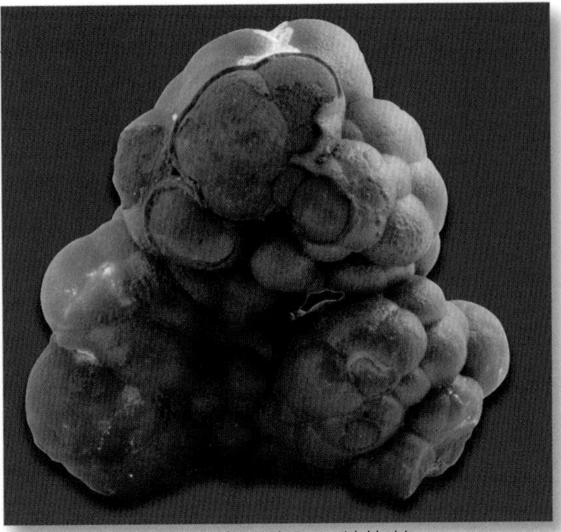

Cassiterite (34598): Botryoidal habit.
Arroyo Carrizal, Guanajuato, Mexico. Width of field of view: 8 cm

Physical Properties

Hardness: Hard (Mohs 6 – 7).

Density: Heavy (7.0 g/cm³).

Breakage: Brittle but tough. It breaks with difficulty and has a poor cleavage and a conchoidal fracture.

Test: Lustre, density, and hardness are important.

Similar Minerals

Chromite (p. 80) has no cleavage and usually has a darker coloured brown streak.

Occurrence

Cassiterite is found associated with igneous granitic rocks, often as veins containing fluorite, topaz, tourmaline, and arsenopyrite. Today many of the tin deposits result from weathered igneous rocks with weather-resistant and heavy cassiterite concentrating in sedimentary gravel (alluvial) deposits (Malaysia, Thailand, Indonesia).

Canada's Best Locality: Big Kalzas Lake, Yukon.

Other Localities: Mono Lake, California, USA; Tasmania, Australia; New South Wales, Australia; Viloco District, La Paz, Bolivia; Minas Gerais, Brazil; Durango, Mexico; Cornwall, England; Horni Slavkov, Czech Republic; Panasqueira, Portugal; Magadan Oblast, Russia; Erongo, Namibia.

Interesting Facts

The mineral name is from the Greek, *kassiteros*, meaning tin, in reference to its composition. Most tin is used in plating steel to prevent corrosion. It has a long history (since 3500 BCE) as one of the main metals in bronze alloys. It is also used in ever-increasing amounts to replace lead in solder for joining electrical wires and plumbing pipes.

Tin's chemical symbol, Sn, is from the Latin, *stannum*, meaning tin.

Cassiterite (50966): Complex crystals with some faces striated. Adamantine lustre. Horni Slavkov, Western Bohemia, Czech Republic. Width of field of view: 6 cm

ANHYDRITE: Calcium sulphate: $CaSO_4$

Anhydrite is a relatively common mineral in sedimentary evaporite deposits – that is, deposits that formed during the evaporation of seawater. Anhydrite can form directly by precipitation from seawater or indirectly by removing water from gypsum through the process of dehydration. Often the two minerals, anhydrite and gypsum, occur together and can be difficult to distinguish.

Areas where sedimentary, evaporite rocks occur often have karst topography. Karst is a land feature marked by sinkholes and abrupt ridges. These surface features form when the soft sedimentary rocks are weathered, producing caverns and underground streams. The land is often dry and barren of most vegetation.

Appearance

Colour: White or colourless to grey or bluish, sometimes blue to violet.

Streak: White to pale grey.

Lustre and transparency: Vitreous or pearly lustre and transparent to translucent.

Habit: Mostly massive or coarse crystalline masses, granular or fibrous. Tabular crystals are rare.

Physical Properties

Hardness: Medium (Mohs 3½).

Density: Medium (3.0 g/cm^3).

Breakage: Brittle with a perfect to good cleavage in three directions perpendicular to each other. An uneven to splintery fracture on the non-cleavage planes.

Test: Colour and hardness.

Similar Minerals

Gypsum (p. 124) is softer (Mohs 2), and has only one perfect cleavage.

Calcite (p. 138) has three perfect cleavages but they are rhombohedral in form and not perpendicular to each other as they are for anhydrite. Calcite fizzes in acid but anhydrite does not.

Occurrence

Anhydrite is found in sedimentary rocks that formed by the evaporation of sea water above 30°C or by the dehydration of gypsum as a result of elevated temperature or pressure. Often anhydrite is associated with gypsum, calcite, and halite. It is less common than gypsum at ground surface as the presence of water favours the formation of gypsum.

Canada's Best Localities: Mauve-coloured, cleavable masses to 15 cm from the Madawaska Mine, Faraday Township, Hastings County, Ontario.

Other Localities: Balmat, Fowler Township, St. Lawrence County, New York, USA; Naica, Mexico; Tuscany, Italy; beautiful pale blue crystals from the Simplon Pass, Switzerland.

Interesting Facts

The mineral name is from the Greek, *an,* meaning without, and *hydros,* meaning water. This name was chosen to emphasize the difference in chemistry from that of gypsum, which is a hydrous calcium sulphate.

Anhydrite and gypsum were first mined in North America in Nova Scotia around 1770; this mining continues today. Anhydrite is used to fill holes and cracks in mines or road tunnels. Gypsum is primarily used for stucco, plaster of Paris, and wallboard.

Anhydrite (38253): Two traces of cleavage visible with the third cleavage as the specimen face. Madawaska Mine, Bancroft, Faraday Tp., Hastings Co., Ontario. Width of field of view: 4.5 cm

GYPSUM: Calcium sulphate hydrate: $CaSO_4 \cdot 2H_2O$

Gypsum is one of the most important industrial minerals with numerous uses in art and building.

Gypsum (84539): Twinned crystals with vitreous lustre. Red River, Winnipeg. Width of field of view: 8 cm

Appearance

Colour: Colourless to white, sometimes grey or yellowish.

Streak: White.

Lustre and transparency: Vitreous lustre with a pearly lustre on cleavage faces, silky lustre if fibrous; transparent to translucent.

Habit: Gypsum has a wide variety of forms. It can occur as crystals: simple single lozenge-shapes, elongate prismatic "swords," or "swallow tail" twins. Most often it is massive, sometimes with a silky, fibrous appearance termed "satin spar."

Physical Properties

Hardness: Soft (Mohs 2).

Density: Light (2.3 g/cm^3).

Breakage: Crystals are flexible but not elastic. Crystals have one perfect cleavage.

Test: Colour and hardness.

Similar Minerals

Calcite (p. 138) is harder (Mohs 3), and has perfect cleavage in three planes (rhombohedral). Anhydrite (p. 122) is harder (Mohs 3), and has perfect cleavage in three perpendicular planes.

Occurrence

It is found in extensive beds resulting from the evaporation of sea water. Due to the insoluble nature of calcium sulphate, gypsum is one of the first minerals to form from marine evaporation.

Canada's Best Localities: Huge deposits of massive gypsum and anhydrite are mined in the Windsor area, Hants County, Nova Scotia. Crystals and satin spar from Willow Creek, Nanton area, Alberta; Swiftcurrent Creek, Moose Jaw area, Saskatchewan; Red River floodway,

Gypsum (40498): Perfect cleavage with pearly lustre. Brunswick #6 Mine, Bathurst, Gloucester Co., New Brunswick. Width of field of view: 14 cm

Winnipeg area, Manitoba; Galetta, Fitzroy Township, Carleton County, Ontario; St. Catharines, Grantham Township, Lincoln County, Ontario; Hillsborough, Albert County, New Brunswick; Ship Cove, Avalon Peninsula, Newfoundland.

Other Localities: Monroe County, New York, USA; Stephen County and Alfalfa County, Oklahoma, USA. Large, spectacular crystals are found in Naica and Santa Eulalia, Chihuahua, Mexico; Aragon, Spain.

Interesting Facts

The mineral name is from the Greek, *gyros*, plaster, alluding to its principal use. As long as nine thousand years ago plaster objects were created in Syria and Anatolia. Today plaster is well known for casting broken limbs and many tons of gypsum a year are used in wallboard and cement.

Gypsum (38447): Single, lozenge-shape crystal with forms, basal pinacoid and two prisms.
Willow Creek, Nanton area, Alberta. Width of specimen: 2 cm

Gypsum (38426): Silky, satin spar variety.
White Basin, Clark Co., Nevada.
Width of specimen: 12 cm

APATITE: Calcium phosphate fluoride: $Ca_5(PO_4)_3F$

Apatite is a group of minerals that includes three species: fluorapatite, $Ca_5(PO_4)_3F$, the most common mineral species of the group, hydroxylapatite; $Ca_5(PO_4)_3(OH)$, and chlorapatite, $Ca_5(PO_4)_3Cl$.

Rarely it can be found as gemmy green crystals that can be faceted or cut as a gemstone, though it is too soft to be used in jewellery.

Apatite mined in Ontario and Quebec during the early part of the 20th century supplied phosphate for the fertilizer industry. This production ceased when large sedimentary phosphate formations were found in other parts of the world.

Appearance

Colour: Usually green but it may be white, yellowish, reddish, brown, or grey.

Streak: White.

Lustre and transparency: Vitreous to resinous lustre and transparent to translucent.

Habit: Often found as crystals with the form of hexagonal prisms and pyramids. May be massive, compact, or granular.

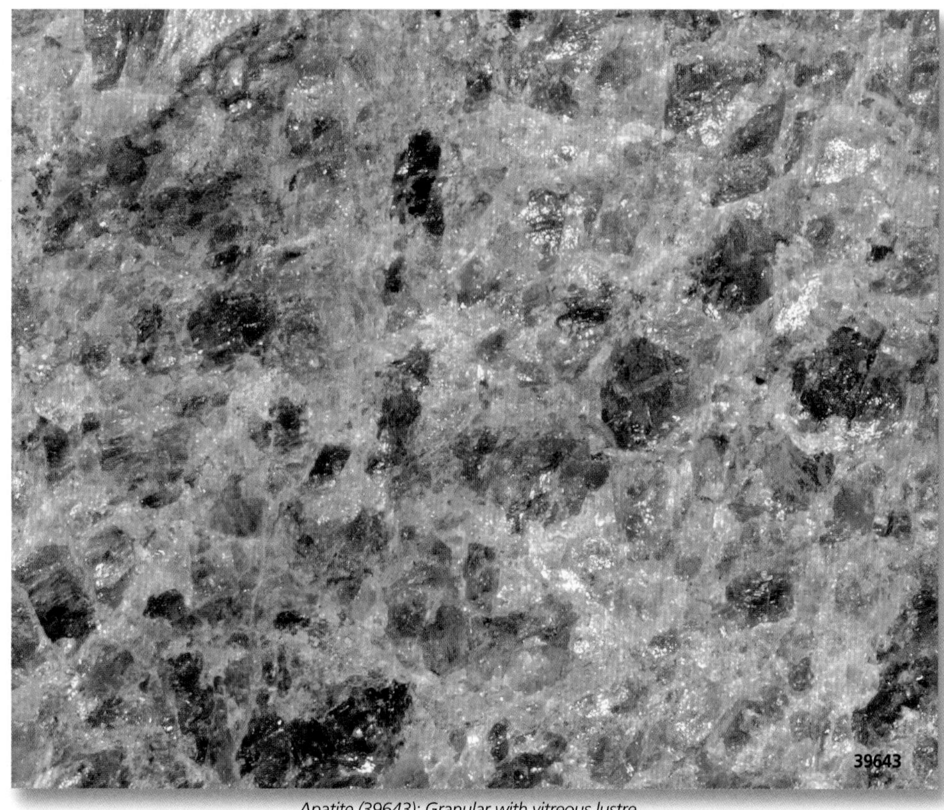

Apatite (39643): Granular with vitreous lustre.
Perth, Drummond Tp., Lanark Co., Ontario. Width of field of view: 7 cm

Apatite (39596): Hexagonal prismatic crystal with hexagonal pyramid termination. Yates Uranium Mine, Otter Lake area, Huddersfield Tp., Pontiac Co., Québec. Width of field of view: 5 cm

Physical Properties

Hardness: Medium hardness (Mohs 5).
Density: Moderately heavy (3.2 g/cm^3).
Breakage: Brittle with no distinct cleavage and an uneven or conchoidal fracture.
Test: Hardness and the lack of cleavage are distinctive.

Similar Minerals

Olivine (peridot variety) (p. 164) is harder (Mohs 6½ – 7), and beryl (p. 184) is also harder (Mohs 7 – 8).

Occurrence

Apatite is found in igneous rocks and metamorphosed limestone (marble).

Canada's Best Localities: Crystals up to 300 kg have been found in the Lake Clear area, Renfrew County, Ontario; light green, transparent crystals from Wilberforce, Monmouth Township, Haliburton County, Ontario; dark blue crystals from Bob's Lake, Verona, Frontenac County, Ontario; crystals to 30 cm from Otter Lake, Pontiac County, Québec.
Other Localities: Minas Gerais, Brazil; Durango, Mexico; Panasqueira, Portugal.

Interesting Facts

The mineral group name is from the Greek, *apatan*, meaning to deceive, because of its similarity to other more valuable minerals.

The main portion of our teeth (dentine), and a minor portion of our bones, has the composition and structure of hydroxylapatite, while the coating or enamel on our teeth has the composition of fluorapatite. This explains the usefulness of dental fluoride treatments. (Teeth are not considered minerals, however, as they are not produced by a geological process.)

Apatite (50513): Hexagonal prism terminated by a small hexagonal pyramid and a pinacoid on pink calcite. Yates Uranium Mine, Otter Lake area, Huddersfield Tp., Pontiac Co., Québec. Crystal length: 6 cm

BARYTE (BARITE)*: Barium sulphate: $Ba(SO_4)$

The first thing you will notice with a baryte specimen is its heft or density. It is quite heavy for a non-metallic mineral, because it contains the heavy element barium. Barium is well past the halfway point in molecular density of all the stable elements. Baryte specimens often have spectacular groups of crystals with bright faces and pleasing colours. Some are found in geodes while others are found as desert roses.

**The original British spelling is "baryte," while "barite" is a common spelling throughout many parts of the world.*

Appearance

Colour: Colourless to white, light blue, yellow, reddish brown.
Streak: White.
Lustre and transparency: Vitreous, resinous, or pearly lustre and transparent to translucent.
Habit: Often as crystals with diverse forms including tabular and prismatic. Crystals are often in rosettes.

Baryte (37553): Tabular, yellow baryte crystals on purple fluorite from the Rock Candy Mine, Grand Forks area, British Columbia. Width of field of view: 3 cm

Baryte (50605): Tabular crystal of baryte with chalcopyrite sprinkled on the face from the Niobec Mine, Chicoutimi Co., Québec. Cleavage traces can be seen at the upper end of the crystal. Width of field of view: 10 cm

Physical Properties

Hardness: Moderate (Mohs 2½ – 3).

Density: Medium (4.5 g/cm^3). This is quite heavy for a non-metallic mineral.

Breakage: Brittle with a distinct cleavage in three directions.

Test: Density, hardness, and cleavage are important physical properties.

Similar Minerals

Aragonite (p. 146) is less dense (2.9 g/cm^3), and has only one perfect cleavage.

Both calcite (p. 138) and dolomite (p. 142) are less dense, have rhombohedral cleavage (three cleavage planes meet at 60º), and fizz in acid.

Occurrence

Baryte is found in hydrothermal deposits, or as veins and lenses in sedimentary limestone rocks. It is found around hydrothermal vents in the oceans.

Baryte (49127): Thick, tabular, rhombic shaped baryte crystals from Touché Claims, Rock River, Yukon. Width of field of view: 7 cm

Baryte (37560): Radiating rosette of baryte crystals from Coppercorp Mine, Algoma District, Ontario. Width of field of view: 8 cm

Canada's Best Localities: Rock Candy Mine, Grand Forks area, British Columbia, has clear yellow platy crystals. Large prismatic crystals up to 40 cm come from the Niobec Mine, Simard Township, Chicoutimi County, Québec; Coppercorp Mine, Algoma District, Ontario, has radiating groups of orange crystals.

Other Localities: Cumbria, England; Morocco; Poland.

Interesting Facts

The name is from the Greek, *barys*, meaning heavy. It was certainly this property of density that earned baryte its name in the 17th century. Because of its density, baryte is used as a weighting agent for petroleum well-drilling mud. This mud or slurry is pumped down the centre of the drill and circulates back up to the surface along the sides of the drill. The weight of the mud must prevent pressurized oil and gas from blowing out the drill hole. A "blow out" is an extremely dangerous part of oil drilling.

Although baryte contains a heavy metal, it is considered non-toxic because of its low solubility. It is used instead of lead compounds to make paint opaque.

As baryte is such a good absorber of X-rays it is also used as a tracer to help detect blockage in the digestive tract; medical procedures include "barium meals" and "barium enemas."

TURQUOISE: Copper aluminium phosphate hydroxyl hydrate: $CuAl_6(PO_4)_4(OH)_8 \cdot 4H_2O$

Turquoise is a distinctive and popular mineral. It is one of the few that contributed its name directly to a colour rather than the mineral named after the colour. Turquoise was being mined for decorative uses before 4000 BCE by the Egyptians in the Sinai Peninsula and as early as 200 BCE by American Indians in the southwestern United States.

*Turquoise (30238): In conglomerate.
Carico Lake Mine, Lander Co., Nevada. Width of specimen: 12 cm*

Appearance

Colour: Most commonly it is "turquoise" – a colour that can be described as pale sky or robin's egg blue. It can also vary from greenish blue to green.

Streak: White with greenish tinge.

Lustre and transparency: Lustre is dull or wax-like and it is translucent to opaque.

Habit: Usually very finely (crypto-) crystalline and massive as nodules and veins in rock. Very rarely crystals are found.

Physical Properties

Hardness: Hard (Mohs 5 – 6).

Density: Medium (2.7 g/cm³).

Breakage: Brittle with a slightly conchoidal fracture.

Test: The colour is so distinctive that the mineral has given its name to the colour.

Similar Minerals

Chrysocolla (p. 216) is softer (Mohs 2 – 4). It is sometimes used to imitate turquoise, which is a much more valuable gemstone.

Malachite (p. 158) is green with no blue and it fizzes in acid.

Occurrence

Turquoise occurs in veins found in altered, near-surface, volcanic rocks associated with limonite and chalcedony.

Canada's Best Localities: Because of glaciations, rocks suitable to the formation of turquoise are not found in Canada.

Other Localities: Santa Fe County, New Mexico, USA; Nye County, Nevada, USA; Campbell County, Virginia, USA; St. Austell, Cornwall, England; Katanga, Democratic Republic of Congo. Some of the earliest known and finest material came from ancient Persia in the Al-Mirsah-Kuh Mountains, northwest of the village of Madan, near Nyshăpùr, Iran.

Interesting Facts

Throughout its long history the mineral has been known by several different names but around the 16th century the French introduced the word, *turquoise*, meaning Turkey. Turkey would have been the trade route for turquoise travelling from Iran and Egypt to Europe. The common usage of turquoise as a term for colour came much later in the mid-19th century. For gem use turquoise is cut as a cabochon with a low-curved upper surface. Turquoise jewellery is so popular today that much of it is imitated ("faked") by chrysocolla or reconstituted from poor quality turquoise by grinding it up, colouring it, and setting it in resin.

Turquoise (40017): In quartz vein. Bisbee, Cochise Co., Arizona. Width of field of view: 7 cm

CALCITE: Calcium carbonate: $Ca(CO_3)$

Calcite is often the first mineral we collect, since it is common and readily forms attractive crystals. It has a great diversity of forms, colours, and localities, which creates problems in identification. The variation in habit has given rise to several variety names that are often very descriptive. For instance, nail head or dogtooth calcite refers to its peculiar but typical forms. Calcite can be easily differentiated from quartz and other similar minerals by using the hardness test.

Appearance

Colour: Colourless, white, pale tints.

Streak: White to grey.

Lustre and transparency: Non-metallic, glassy lustre, translucent to transparent.

Habit: Massive to granular, crystals pyramidal, tabular, needle-like. The most common crystal form, rhombohedron, consists of six faces (three up and three down) each the shape of a rhomb or diamond.

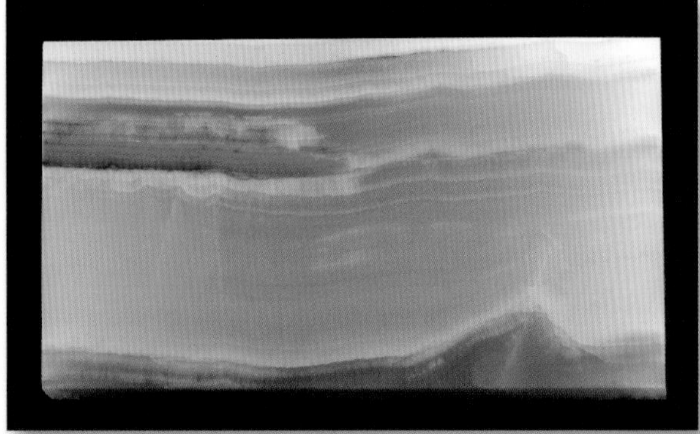

Calcite (35576): Mexican onyx variety name. Mexico Width of polished plate: 10 cm

Physical Properties

Hardness: Medium (Mohs 3).

Density: Medium (2.7 g/cm^3).

Breakage: Brittle with good rhombohedral cleavage.

Test: The crushed mineral will fizz when several drops of vinegar are added.

Calcite (35665): Three perfect cleavages making a rhombohedron. Note internal cleavage planes. Iceland spar variety. Creel, Chihuahua, Mexico. Width of specimen: 7 cm

Similar Minerals

Feldspar (p. 226) and quartz (p. 218) are harder.

Fluorite (p. 106) is usually more strongly coloured and heavier. Unlike calcite, fluorite has cubic cleavage.

Dolomite (p. 142) fizzes much less or not at all in a weak acid such as vinegar.

Aragonite (p. 146) has only one cleavage.

Occurrence

This very common mineral is found in sedimentary (limestone p. 276) and metamorphic

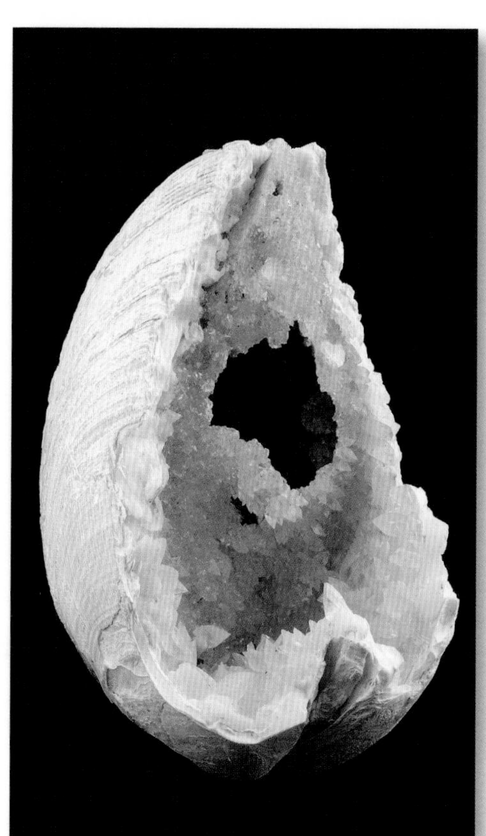

Calcite (35747): Fine pyramidal crystals filling a bivalve fossil shell. Hampton, York Co., Virginia. Width of field of view: 9 cm

Calcite (58446): Poker chip variety showing a short hexagonal prism terminated by a pinacoid. St. Andreasburg, Lower Saxony, Germany. Width of field of view: 9 cm

Calcite (84408): Complex rhombohedral crystals with vitreous lustre. Clear outer crystal with an earlier formed orange inner crystal called a phantom. Grant Quarry, Greely, Osgoode Tp., Carleton Co., Ontario. Width of field of view: 4 cm

Calcite (30184): Rhombohedral crystal, layered overgrowth. Note rhombohedral cleavage. Bluebell Mine, Riondel, Kootenay District, British Columbia. Width of crystal: 6 cm

(marble p. 294) rocks. Recent discoveries indicate that calcite is sometimes formed deep inside the earth in igneous rocks such as carbonatite.

Canada's Best Localities: Large pink or white rhombohedral crystals from the Bluebell Mine, Riondell, Kootenay District, British Columbia; bowtie crystal groups from the George McLeod Mine, Chabanel Township, Algoma District, Ontario; double-terminated crystals from Coppercorp Mine, Algoma District, Ontario; large blocky crystals from the Long Lac Mine, Olden Township, Frontenac County, Ontario; blocky crystals to 30 cm from Lyndhurst, Leeds and Lansdowne Townships, Leeds County, Ontario; hematite-coated crystals from the

Calcite (53750): Vitreous lustre, rough, scalenohedral crystal. Nanisivik Mine, Nanisivik, Baffin Island, Nunavut. Width of field of view: 4.5 cm

Madawaska Mine, Faraday Township, Hastings County, Ontario; flat, triangular crystals from Niobec Mine, Simard Township, Chicoutimi County, Québec; twinned crystals from Baie Comeau, La Fleche Township, Saguenay County, Québec.

Other Localities: Hardin County, Illinois, USA; Jasper County, Missouri, USA; Ottawa County, Oklahoma, USA; Elmwood, Smith County, Tennessee, USA; Chihuahua, Mexico; San Luis Potosi, Mexico; Derbyshire, England; Cumbria, England; Cornwall, England; St. Gallen, Switzerland; Tsumeb, Namibia.

Interesting Facts

Double refraction is observed in fragments of fractured calcite. If the fragment is placed on a printed page it doubles the image. Calcite fluoresces under ultraviolet light when trace amounts of magnesium are present.

Calcite is used primarily in the manufacture of mortar and plaster. The clear, colourless variety, Iceland spar, is used in optical instruments. The term "Mexican onyx" should not be used as "onyx" is actually a black coloured variety of quartz. Unfortunately the term is widely used in reference to a large number of decorative items, often dyed, that have been made from calcite and, or aragonite.

Calcite (43711): Stalactitic habit.
Carter Co., Montana.
Width of field of view: 15 cm

DOLOMITE: Calcium, magnesium carbonate: $CaMg(CO_3)_2$

Dolomite (36252): Rounded, rhombohedral crystals with cleavage planes visible. Nova Lima, Minas Gerais, Brazil. Width of specimen: 12 cm

Dolomite is a common mineral that can be difficult to distinguish from calcite, another common carbonate mineral. Dolomite is an important mineral because of its many uses. It is the major component of two common rocks, fine-grained dolostone and coarser-grained, dolomitic marble.

Appearance

Colour: White or grey and sometimes there are tints of pink, green or brown.

Streak: White

Lustre and transparency: Vitreous to pearly lustre and transparent to translucent.

Habit: Often as rhombic or diamond-shaped crystals with curved faces. If the crystals are made up of several individuals the curved faces may be saddle-shaped.

Physical Properties

Hardness: Medium (Mohs 3-4).

Density: Medium (2.9 g/cm^3).

Breakage: Brittle with a distinct cleavage on three planes giving a rhombohedral shape and an uneven, sub-conchoidal fracture on the non-cleavage planes.

Test: Curved crystal faces. It fizzes only in warm or strong acid. In a weak acid, like vinegar, there is almost no fizz.

Dolomite (49860): Marked rhombohedral cleavage. Dusting of chalcopyrite. Asarco Aquarius Mine, Timmins area, Macklem Tp., Cochrane District, Ontario. Length of field of view: 5 cm

Similar Minerals

Calcite (p. 138) fizzes a lot in acid while dolomite fizzes much less or not at all in a weak acid such as vinegar.

Occurrence

Dolomite is found in sedimentary rocks of marine origin, much like calcite. It is also found in hydrothermal veins commonly associated with fluorite and baryte.

Canada's Best Localities: Pale orange-coloured crystals on red quartz from Rabbit Lake, Saskatchewan; Saint Catharines, Lincoln County, Ontario; Trudeau Quarry, Vaudreuil County, Québec; pink saddle-shaped crystals from Saint-Eustache, Québec; Gasparine, Chateauguay County, Québec.

Other Localities: Fine crystals from Traversella and Brosso, Piedmont, Italy and from Eugui, Navarra Province, Spain; Belo Horizonte, Minas Gerais and Bahia, Burmado District, Brazil.

Interesting Facts

In 1794 this mineral was named for the French geologist, Déodat Gratet de Dolomieu, who first described this mineral from his studies of the Alps.

Dolomite is a source of magnesium (Mg) metal and magnesia (MgO). The metal is lightweight, hence its applications in aircraft, spacecraft, automobiles and portable tools.

Dolomite (43605): Curved, pink crystals with younger, colourless, complex crystals on top. St.-Eustache Quarry, St.-Eustache, Deux-Montagnes Co., Québec. Width of field of view: 5 cm

ARAGONITE: Calcium carbonate: $Ca(CO_3)$

Aragonite has the same chemical composition as calcite but the atomic structure of the two minerals is quite different. Hence, the two minerals are polymorphs of each other. Aragonite is less common than calcite but it is an important mineral that is found in a variety of geological environments. Its most important occurrence is as animal secretions in pearls and sea shells. Because of this biological origin, there is some debate as to whether this form of aragonite is a true mineral.

Appearance

Colour: Colour is quite variable; normally white but it can also be grey, yellow, green, blue, violet, reddish, or brown.

Streak: White.

Lustre and transparency: Vitreous lustre and translucent.

Habit: Often massive, can be stalactitic, as crusts, or coral-like. Crystals can be prismatic to needle-like. Sometimes crystals are twinned, forming hexagonal-shaped prisms.

Aragonite (35984): Hexagonal prism is twinned. Amarillo, Potter Co., Texas. Width of field of view: 4 cm

Physical Properties

Hardness: Moderate (Mohs 3-4).

Density: Medium (2.9 g/cm^3).

Breakage: Brittle with one distinct cleavage and a conchoidal fracture on the non-cleavage planes.

Test: Aragonite fizzes in dilute acid.

Similar Minerals

Calcite (p. 138) has perfect cleavage in three planes, not just one like aragonite.

Dolomite (p. 142) does not fizz in weak acid.

Occurrence

Aragonite is found in sedimentary rocks. It forms around hot springs and in caves as stalactites and stalagmites.

Aragonite (32958): Crust on rock surface.
Jeffrey Mine, Asbestos, Shipton Tp., Richmond Co., Québec.
Width of field of view: 15 cm

Canada's Best Localities: Sprays of acicular crystals to several centimetres have been found at Thetford Mines, Mégantic County, Québec; crystals to 8 cm from Caland Iron Mine, Rainy River District, Ontario.

Other Localities: Amarillo, Potter County, Texas, USA; Katharine Mine, Madison County, Missouri, USA; near Carlsbad, New Mexico, USA; Broken Hill, New South Wales, Australia; Molina de Aragon, Guadalajara Province, Spain; Styria Province, Austria; Herrengrund, Czech Republic; Agrigento, Sicily, Italy; Cumbria, England.

Interesting Facts

The mineral name is for the Aragon region of Spain, where it was first described in 1767.

Aragonite normally forms at higher pressures than calcite but interestingly it is aragonite that forms as pearls and the shells of some sea animals. The reason for this low temperature and low pressure formation of aragonite is unknown.

Often we see the term Mexican onyx applied to a wide variety of decorative objects such as tables, book ends, chess pieces, ashtrays, and jewellery. It can occur in several muted colours such as white, tan, or cream-yellow but it is often dyed in more vivid colours such as blue, green, and red. This material is travertine, a rock composed of calcareous deposits from ground water. It is a mixture of calcite and aragonite.

Aragonite (35969): Hexagonal twin. Sicily, Agrigento District, Italy. Width of field of view: 7 cm

Aragonite (39373): Spray of translucent needles. Thetford Mines area, L'Amiante, Thetford Tp., Mégantic Co., Québec. Width of field of view: 7 cm

SIDERITE: Iron carbonate: $Fe(CO_3)$

Siderite has the same crystal structure as calcite, with iron atoms replacing calcium atoms. With the same crystal structure, its crystal form and cleavage are similar to those of calcite, but siderite has major differences in colour, density, and hardness.

Appearance

Colour: Most commonly light to dark brown, sometimes tinted greenish or reddish.

Streak: White.

Lustre and transparency: Vitreous lustre and translucent.

Habit: Often as rhombohedral crystals that sometimes have curved faces. It may also be massive, stalactitic, or rounded.

Siderite (36842): Rhombohedral crystal on albite.
Demix & Poudrette Quarries, Mt. St.-Hilaire, Rouville Co., Québec.
Width of field of view: 11 cm

Physical Properties

Hardness: Moderate (Mohs 4).

Density: Medium (4.0 g/cm^3).

Breakage: Brittle with a perfect cleavage parallel to the rhombohedral planes.

Test: Colour, hardness, and density are important. Fizzes in dilute acid.

Similar Minerals

Calcite (p. 138) is usually colourless or white and not brown. Calcite is less dense (2.7 g/cm^3) and softer (Mohs 3) than siderite.

Dolomite (p. 142) can resemble siderite as it has a similar density (2.9 g/cm^3) and hardness (Mohs 3 – 4). A careful acid test can differentiate them, as siderite effervesces more in dilute acid than dolomite.

Occurrence

Siderite is found most often in sedimentary environments as bedded layers within clay, shale, or coal. These deposits may then be metamorphosed giving a more crystalline texture to the rock. Rarer occurrences of siderite include hydrothermal deposits and granite and nepheline syenite pegmatites.

Canada's Best Localities: Large rhombohedral crystals up to 15 cm have been found at Mont Saint-Hilaire, Rouville County, Québec; Rapid Creek, Yukon.

Other Localities: Potosi, Bolivia; Minas Gerais, Brazil; Cornwall, England; Harz Mountains, Germany; Bindt, Slovakia; Baia Sprie, Romania.

Interesting Facts

The name siderite is from the Greek, *sideros*, meaning iron.

Siderite is a minor but important mining source for iron. At the Helen Mine near Wawa, Ontario, banded siderite iron ore was mined profitably for 60 years (1937 – 1997). Siderite was also mined at the Rudnany ore field, Slovakia.

There are enormous iron formations around the globe that contain siderite; Lake Superior region of North America; Labrador Trough, Canada; Hamersley Range, Western Australia; and the Transvaal Supergroup of South Africa. A vast sea that covered much of the earth prior to 3.5 billion years ago contained large amounts of iron that had derived from the weathering of iron-rich rocks in an atmosphere low in oxygen and high in methane. With the evolution of the atmosphere, methane decreased, allowing life to begin. Increased oxygen and carbon dioxide, released through photosynthesis by primitive blue-green algae, precipitated iron as iron carbonate or siderite. Further increases in oxygen favoured the precipitation of goethite, hematite, and magnetite.

Siderite (55154): Stepped, rhombohedral crystal. Morro Velho Mine, Nova Lima, Minas Gerais, Brazil. Width of field of view: 4 cm

RHODOCHROSITE: Manganese carbonate: $Mn(CO_3)$

Rhodochrosite is one of the very few pink minerals, so it is rather easy to identify. It comes in several forms and shades of colour, making it popular among mineral collectors. It also makes attractive jewellery but it is not durable due to its softness. Be careful not to confuse the name with rhodonite which is manganese silicate.

Appearance

Colour: Red through shades of pink to brown to white. Often patterned with white bands of calcite.

Streak: White.

Lustre and transparency: Vitreous lustre and transparent to translucent.

Habit: Several habits such as massive, rounded, or botryoidal, stalactitic and fine rhombohedral crystals (like calcite).

Rhodochrosite (48247): Polished section of a stalactite. Catamarca, Argentina. Width of specimen: 10 cm

Rhodochrosite (37067): Rhombohedral crystals.
Demix & Poudrette Quarries, Mt. St.-Hilaire,
Rouville Co., Québec.
Width of specimen: 3 cm

Physical Properties

Hardness: Moderate (Mohs 3 – 4).

Density: Medium (3.7 g/cm^3).

Breakage: Brittle with a good rhombohedral cleavage and an uneven fracture on the non-cleavage planes.

Test: Colour and hardness. Also, it fizzes in acid.

Similar Minerals

Rhodonite (p. 202) is harder (Mohs 6), has a cleavage close to 90°, and does not fizz in acid.

Occurrence

Rhodochrosite is found in veins that have formed from hot hydrothermal solutions.

Canada's Best Localities: Mount Saint-Hilaire, Rouville County, Québec; Bluebell Mine, Kootenay District, British Columbia.

Other Localities: Crystals from Park County, Colorado, USA; Iron County, Wisconsin, USA; Peru and the Cape Province, Republic of South Africa; banded, stalactitic forms from Catamarca, Argentina; Stratoni, Greece.

Interesting Facts

The mineral name is from Greek, *rhodon*, rose, plus *chrosis*, a colouring. Rhodochrosite is of minor importance as a source of manganese. Its main use is in jewellery as beads and pendants.

Rhodochrosite (37078): Rhombohedral cleavage.
Demix & Poudrette Quarries, Mt. St.-Hilaire,
Rouville Co., Québec.
Width of field of view: 7 cm

AZURITE: Copper carbonate hydroxyl: $Cu_3(CO_3)_2(OH)_2$

The blue of azurite is unforgettable and specimens are highly prized. Azurite was used as a source of blue pigment from the time of Egypt's Fourth Dynasty until the 19th century. It was the most important pigment during the Middle Ages and Renaissance until the synthetic pigment, Prussian blue, began to replace it in the early 1700s.

Like its more common close associate, malachite, azurite is only of medium hardness, and can be used only for jewellery that is not treated roughly, such as pendants. Sometimes malachite and azurite are found together forming beautiful intergrown patterns.

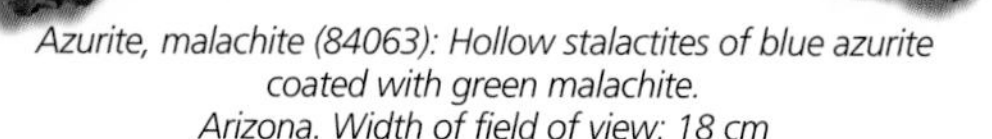

Azurite, malachite (84063): Hollow stalactites of blue azurite coated with green malachite. Arizona. Width of field of view: 18 cm

Appearance

Colour: Dark blue, almost blue-black in some specimens.

Streak: Light blue.

Lustre and transparency: Vitreous lustre on crystal and duller lustre on massive specimens. It is translucent to opaque.

Habit: Varies from tabular to stubby or elongate-prismatic crystals. It is commonly massive, with botryoidal, stalactitic, or radiating habits.

Physical Properties

Hardness: Medium (Mohs 3 – 4).

Density: Moderate (3.8 g/cm^3).

Breakage: Brittle with a distinct cleavage and a conchoidal fracture on the non-cleavage planes.

Test: Colour and hardness. It fizzes easily in strong acid such as hydrochloric.

Similar Minerals

Sodalite (p. 234) is much harder (Mohs 6) and has no distinct cleavage.

Lazurite (p. 236) is harder (Mohs 5 – 6) and has no distinct cleavage.

Occurrence

It is found in the altered zone (oxidation zone) of copper deposits commonly associated with malachite, chrysocolla, cuprite, and calcite.

Canada's Best Localities: Small patches of bright, microscopic crystals from Highland Valley Copper Mine, Kamloops District, British Columbia; Pueblo Mine, Whitehorse District, Yukon.

Other Localities: Bisbee, Cochise County, Arizona, USA; Toussit, Morocco; Tsumeb, Namibia; Katanga, Democratic Republic of Congo; Chessy, Rhone Department, France; Moonta, South Australia, Australia.

Interesting Facts

The mineral name is derived from the Persian, *lazhward*, meaning blue colour.

The chemical formulae of malachite, $Cu_2(CO_3)(OH)_2$, and azurite, $Cu_3(CO_3)_2(OH)_2$, are very similar. In exceptional cases malachite can be found replacing azurite crystals; this is called a pseudomorph or "false form." The pseudomorphic replacement is a result of oxidation of azurite to produce malachite. This process has altered the colour of some paintings produced centuries ago.

Azurite (48296): Prismatic crystals. Touissit Mine, Oujda, Touissite, Morocco. Width of field of view: 4 cm

MALACHITE: Copper carbonate hydroxyl: $Cu_2(CO_3)(OH)_2$

The green colour typical of copper-bearing minerals is most evident in malachite. Its banded light and dark green designs are so distinctive that malachite is one of few minerals easily recognized by the general public. Unfortunately, it is too soft, like all carbonate minerals, to produce durable jewellery.

Appearance

Colour: Light to dark – almost blackish green. In masses it often has a beautiful banded pattern.

Streak: Light green.

Lustre and transparency: The lustre varies from vitreous to adamantine on crystal faces, silky in fibrous forms, and earthy in massive forms. It is translucent to opaque.

Habit: Often it is massive with globular or stalactitic shapes. Malachite crystals are rare; they vary from needles to equant (equal dimensions in all three directions), tabular forms.

Malachite (32015): Radiating needles. Bristol Silver Mine, Pioche, Lincoln Co., Nevada. Width of field of view: 5 cm

Physical Properties

Hardness: Moderate (Mohs 3½ – 4).

Density: Medium (4.0 g/cm^3).

Breakage: Brittle with a distinct cleavage and an uneven fracture on massive forms.

Test: Colour and hardness. It fizzes in strong acid such as hydrochloric.

Similar Minerals

Chrysocolla (p. 216) is lighter in colour and it does not fizz in strong acid.

Occurrence

It is found in the near surface or altered zone (oxidation zone) of copper deposits, and malachite is commonly associated with chrysocolla, cuprite, azurite, and calcite.

Canada's Best Localities: Jellicoe, Thunder Bay District, Ontario; Craigmont Mine, Kamloops District, British Columbia; Pueblo Mine, Whitehorse District, Yukon.

Other Localities: Bisbee, Cochise County, Arizona, USA; Nizhiny Tagil, Ural Mountains, Russia; Kakanda, Katanga Province, Democratic Republic of Congo; Tsumeb, Namibia.

Interesting Facts

The mineral name is derived from the Greek *moloche*, or mallow, a dark green plant originally from Asia and Europe that has now spread as a weed all over the world.

Malachite was used as a green pigment on Egyptian tomb paintings of the Fourth Dynasty (approximately 2600 – 2500 BCE) and later in 15th and 16th century Europe. A thin veneer of malachite adorns massive columns and urns in the "Malachite Room" of the Hermitage Museum of St. Petersburg, Russia.

Albertus Magnus, the great 13th century German philosopher and alchemist, wrote that the mineral malachite protects the wearer against harm, and could be used to guard babies' cradles.

Malachite (37233): Nodular with concentric banding. Nizhniy Tagil, Russia. Width of field of view: 16 cm

BORAX: Sodium borate hydrate: $Na_2B_4O_5(OH)_4 \cdot 8H_2O$

This mineral became celebrated through the product Twenty Mule Team Borax, a household cleaner and disinfectant. Borax is a mineral found in large quantities but in very few places. It has a special mode of occurrence.

Appearance

Colour: Colourless but it readily loses water and alters to white, grey, or yellow.
Streak: White.
Lustre and transparency: Vitreous to dull or chalky (when it dehydrates) lustre and translucent.
Habit: Often as prismatic-shaped crystals.

Physical Properties

Hardness: Soft (Mohs 2 – 2½).
Density: Light (1.7 g/cm^3).
Breakage: Brittle with a distinct cleavage and a conchoidal fracture on the non-cleavage planes.
Test: Colour, powdery alteration, and hardness.

Similar Minerals

Ulexite (p. 162) needs hot water to dissolve while borax is soluble even in cold water. Ulexite is more fibrous in habit while borax is prismatic.

Occurrence

Borax is found in sedimentary deposits formed by evaporation. It is also found associated with hot springs. Borax is often found with ulexite.

Canada's Best Localities: Although borax is unknown in Canada, there are many borate minerals found in the potash deposits of Sussex, Kings County, New Brunswick.
Other Localities: Found in many salt lakes in Nevada, and California.

Interesting Facts

The mineral has been known since antiquity and the name derives from Persian, *burah*, meaning white.

Like ulexite, borax is a major source of the element boron. This exotic element finds wide usage in medicine, glass, ceramic glazes, new-age materials in jets, and rocket fuels.

At present there is much interest in the use of boron as a safe medium by which hydrogen fuel can be stored. Hydrogen as fuel could one day be an important replacement of gasoline.

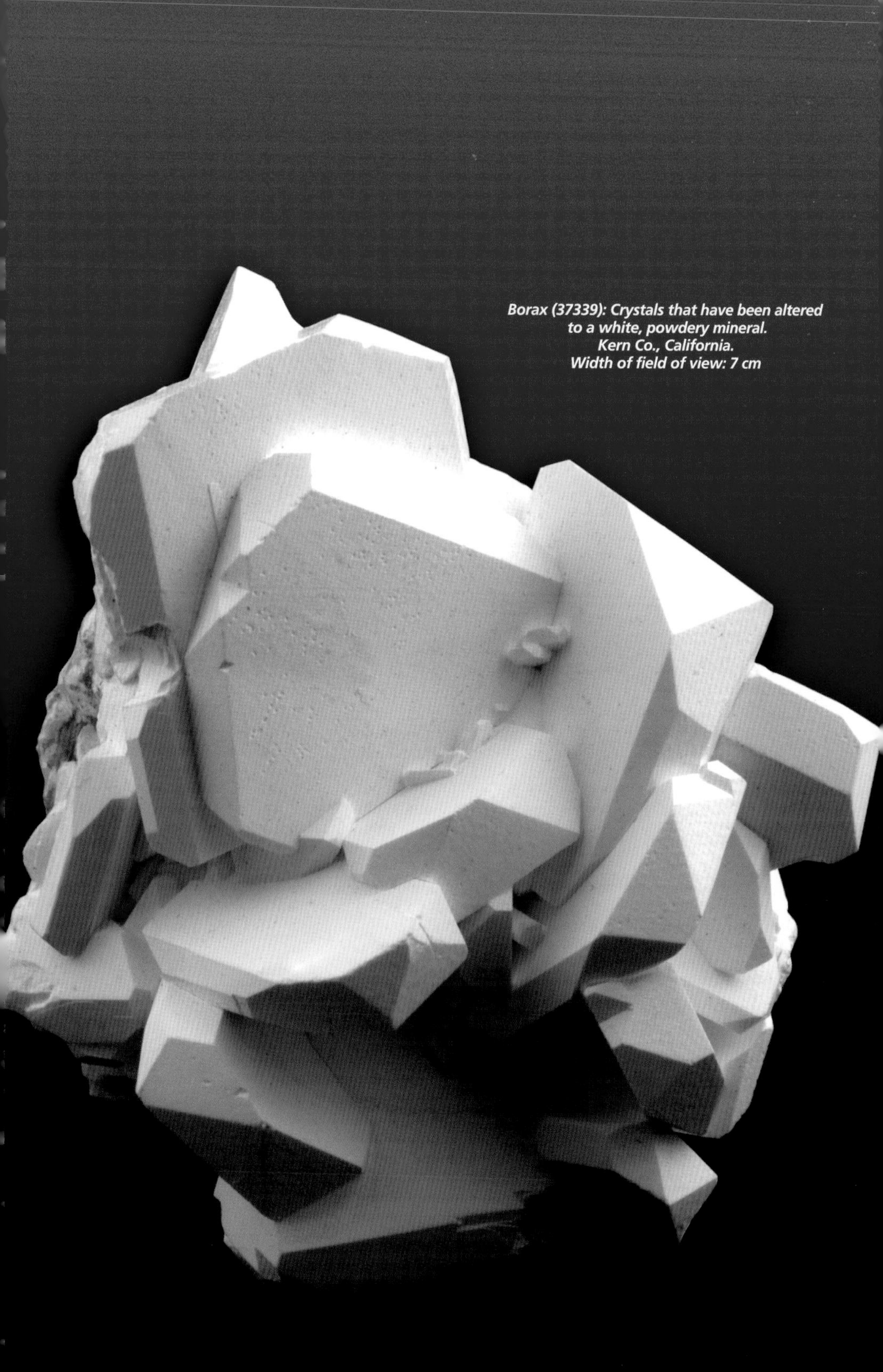

Borax (37339): Crystals that have been altered to a white, powdery mineral. Kern Co., California. Width of field of view: 7 cm

ULEXITE: Sodium calcium borate hydrate: $NaCaB_5O_6(OH)_6 \cdot 5H_2O$

Ulexite (37343); Silky, fibrous "TV rock variety."
Boron, Kern Co., California. Width of specimen: 6 cm

Ulexite has come to be known as TV rock due to its interesting and valuable optical properties. With flat polished faces perpendicular to its fibres, a specimen will display an image of whatever is adjacent to the opposite side. This fibre-optic effect results from the polarization of light and the phenomenon is widely used in many types of visual and data transmission today.

Appearance

Colour: White, grey, or colourless.
Streak: White.
Lustre and transparency: Vitreous or silky lustre and transparent to translucent.
Habit: Often as fibres, either in radiating tufts or compact.

Physical Properties

Hardness: Soft (Mohs 2). Hardness is difficult to measure because of the fibrous habit.
Density: Light (2.0 g/cm^3).
Breakage: Brittle with a distinct cleavage.
Test: Dissolves in hot water.

Similar Minerals

Borax (p. 160) is soluble in cold water and is prismatic in shape.

Occurrence

Ulexite is found in sedimentary deposits formed by the evaporation of lakes or marshes in arid climates. The boron is from volcanic activity. Ulexite is often associated with borax.

Canada's Best Locality: Found in the potash deposits of Sussex, Kings County, New Brunswick.

Other Localities: Found in many salt lakes in Nevada, and Kern County and Inyo County, California, USA.

Interesting Facts

The mineral is named after George Ludwig Ulex, a German chemist who was the first to correctly analyze it.

Ulexite is an important ore of boron. Boron finds a wide variety of uses in medicine, rocket fuels, ceramic glazes, and disinfectants. Synthetic boron nitride is as hard as diamond and is used as an abrasive.

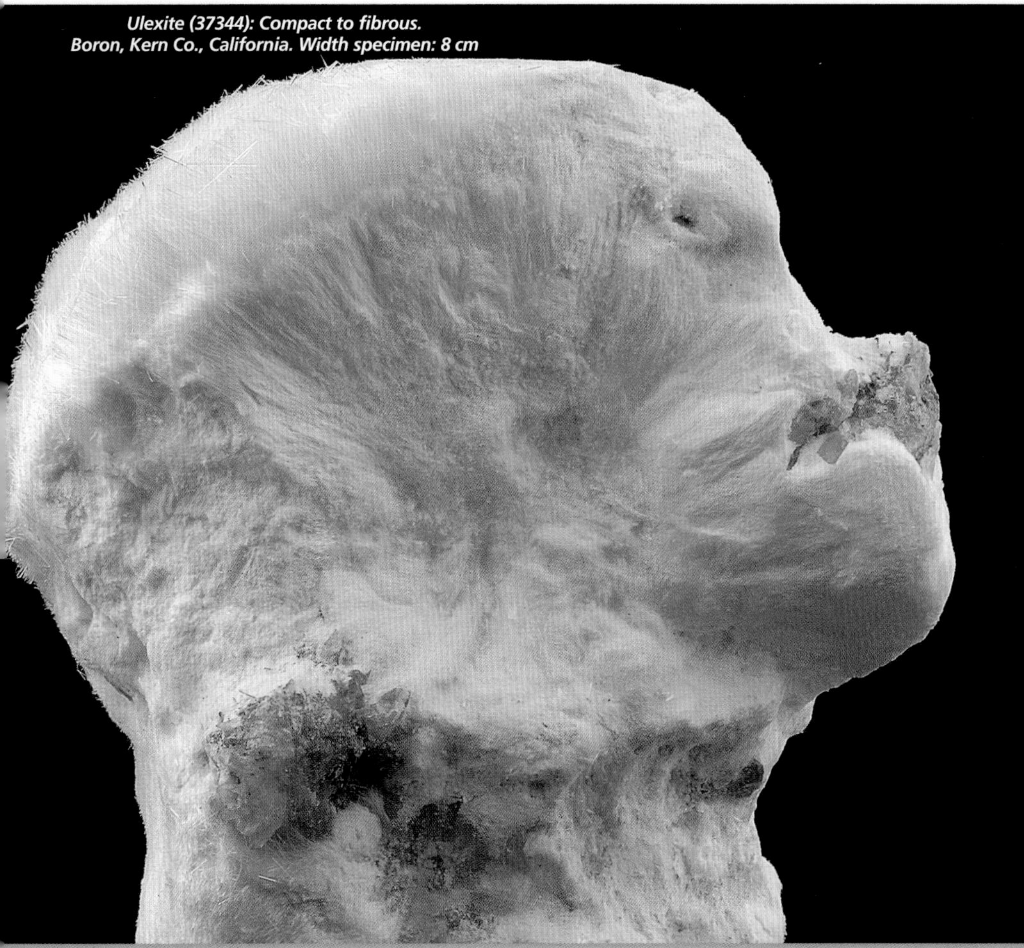

Ulexite (37344): Compact to fibrous. Boron, Kern Co., California. Width specimen: 8 cm

OLIVINE: Magnesium, iron silicate: $(Mg,Fe)_2SiO_4$

Olivine (37458): Tabular crystal with a large basal pinacoid and small pyramid forms in pale pink calcite. Parker Mine, Notre-Dame-du-Laus, Bigelow Tp., Labelle Co., Québec. Width of field of view: 6 cm

Olivine is the name of a chemical composition series that extends from pure forsterite, Mg_2SiO_4 to pure fayalite, Fe_2SiO_4. Forsterite is much more common than fayalite. It is difficult to tell the two minerals apart and most specimens contain both iron and magnesium, making them intermediate in composition between forsterite and fayalite. For these reasons it is appropriate to use the series name, olivine, for identification purposes. Although olivine is not a common mineral in the earth's crust, it is very common in the mantle, the layer below the crust.

Appearance

Colour: Yellow-green, olive green, brown to black; the darker colours are due to increased iron content.

Streak: Olivine is too hard to take a streak but the powder is white.

Lustre and transparency: Vitreous lustre and transparent to translucent.

Habit: Often as rounded grains or massive; rare crystals are thick, tabular with wedge-shaped termination.

Physical Properties

Hardness: Hard (Mohs 6-7).
Density: Medium (3.3 g/cm^3 for forsterite to 4.4 g/cm^3 for fayalite).
Breakage: Brittle with no distinct cleavage and a conchoidal fracture.
Test: Colour and hardness are important. Occurs in dark-coloured magmatic rocks.

Similar Minerals

Pyroxene (p. 192) and amphibole (p. 196) have two good cleavages.

Occurrence

Olivine is found in magmatic rocks that have very little silica (SiO_2), such as basalt, peridotite, and dunite. It is also found in metamorphosed impure limestones and meteorites.

Canada's Best Localities: Soper Lake, Baffin Island, Nunavut; Lightning Peak, near Cherryville, Yale District, British Columbia; Parker Mine, Bigelow Township, Labelle County, Québec.

Other Localities: Near Globe, Arizona, USA; Dona Ana County, New Mexico, USA; Smrci, Czech Republic; Canary Islands, Spain; Kashmir, Pakistan. Beautiful, gemmy green crystals at St. Johns Island, Red Sea, Egypt.

Olivine (41718): Pale coloured, peridot variety in basalt. Lightning Peak, Osoyoos Division, Yale District, British Columbia. Width of field of view: 8 cm

Interesting Facts

The mineral name olivine refers to its olive-green colour.

The gemmy green variety of olivine (forsterite in this case) is known as peridot (pronounced per-i-doe). This gem is readily affordable and has been in use since Roman times.

A whole series of synthetic forsterite lasers have been produced for a variety of applications such as micromachining, eye surgery, in vivo imaging of cells, and other biological applications.

GARNET GROUP

Iron, magnesium, manganese, calcium, aluminum silicate: $(Fe,Mg,Mn,Ca)_3(Al,Fe)_2(SiO_4)_3$

Garnet comprises a group of minerals of 18 different species having different chemical compositions. They are relatively common in metamorphic rocks and are readily identified by their characteristic crystal shape. In some cases one species may be differentiated from another by colour while in other instances the mode of occurrence helps.

The six most common garnets:

Almandine $Fe_3Al_2(SiO_4)_3$; deep red, reddish brown to almost black; common in metamorphic rocks like schist (p. 286); also called carbuncle now and in ancient times.

Pyrope $Mg_3Al_2(SiO_4)_3$; ruby red, purplish red (gem variety rhodolite), almost black; a rarer garnet mineral found in peridotite; used as a possible "indicator" mineral in searching for diamonds.

Spessartine $Mn_3Al_2(SiO_4)_3$; violet-red or even orange-yellow; a rare garnet mineral found in magmatic rocks such as granite (p. 246) and rhyolite (p. 256).

Andradite $Ca_3Fe_2(SiO_4)_3$; yellow, red, brown, green, or black; in contact metamorphosed limestone (marble p. 294)

Grossular $Ca_3Al_2(SiO_4)_3$; yellow, orange-yellow, green; a fairly rare garnet found in contact metamorphosed limestone (marble p. 294) with vesuvianite and diopside.

Uvarovite $Ca_3Cr_2(SiO_4)_3$; bright green; a rare garnet found in metamorphic marbles (p. 294) and serpentinites (p. 298) as well as in magmatic rocks such as peridotite and kimberlite.

Garnet (32026): Perfect dodecahedral grossular garnets. The green colour is derived from the element chromium and in some crystals you can see a dark core which is chromite.
Jeffrey Mine, Asbestos, Shipton Tp., Richmond Co., Québec. Width of field of view: 3 cm

Garnet (31297): Broken, rounded almandine crystals in cream coloured feldspar and quartz. Garnets do not have cleavage. Southeast of Soper Lake, Baffin Island, Nunavut. Width of specimen: 8 cm

Appearance

Colour: Red, red-brown to almost black are the common colours but it can also be colourless, white, grey, pink, yellow-orange, and green.

Streak: Too hard to streak but garnet powdered is white.

Lustre and transparency: Vitreous, sometimes resinous lustre and transparent to translucent.

Habit: Often as crystals that have equal dimensions in all three directions (equi-dimensional). The most common crystal forms are the rhombic (diamond-shaped) dodecahedron (12 faces) and the trapezohedron (12 kite-shaped faces).

Sometimes the crystals are rounded to appear as spheres.

Physical Properties

Hardness: Hard (Mohs 7).

Density: Medium (3.5 – 4.3 g/cm^3).

Breakage: Brittle with no distinct cleavage and an uneven or conchoidal fracture.

Test: Colour, crystal shape, and hardness are important.

Garnet (45909): Almandine crystal on mica, garnet gneiss. The complex crystal has three forms evident, rhombic, dodecahedron (diamond-shaped face) and trapezohedrona, (kite-shaped face with six edges). Wrangell Island, Wrangell, Alaska. Width of field of view: 5 cm

Similar Minerals

Garnet species are difficult to distinguish from each other without chemical analyses but colour and occurrence help.

Vesuvianite (p. 182) can be a similar colour and lustre but it has elongate, square-prismatic crystals while garnet is equi-dimensional in shape.

Occurrence

Garnet is found in metamorphic rocks like mica schist (p. 286) or marble (p. 294).

Canada's Best Localities: Almandine from Cape Dorset, Baffin Island, Nunavut; River Valley, Dana Township, Nipissing District, Ontario. Grossular from Raft River, Kamloops District, British Columbia; Coe Hill, Wollaston Township, Hastings County, Ontario. Colourless, green, and orange crystals are found in the Jeffrey Mine, Asbestos, Québec. Andradite from Marmoraton Iron Mine, Marmora Township, Hastings County, Ontario.

Garnet (32536): Dodecahedral almandine crystals on mica schist. Austria. Width of field of view: 16 cm

Garnet (32765): Grossular garnet from the Jeffrey Mine, Asbestos, Shipton Tp., Richmond Co., Quebec. Each crystal has the form of a rhombic dodecahedron, rhombic (a rhomb) is diamond shaped and dodecahedron means twelve faces. Width of field of view: 4 cm

Other Localities: Almandine from Archipelago, Alaska, USA; Chaffee County, Colorado, USA; Tyrol, Austria; Thackaringo, New South Wales, Australia; Karoi, Zimbabwe. Spessartine from Thomas Range, Juab County, Utah; Broken Hill, New South Wales, Australia; Rio Grande do Norte, Brazil. Grossular from Eden Mills, Vermont, USA; Lake Jaco, Mexico. Andradite from Stanley Butte, Arizona, USA; San Benito County, California, USA; Hidalgo, Zacatecas, and Chihuahua, Mexico; Val Malenco, Italy. Uvarovite from Outokumpu, Finland; Permskaya Oblast', Russia.

Interesting Facts

The mineral name garnet is known from antiquity. It is from the Latin word, *granatum*, meaning seed. It most likely refers to *malum granatum*, the pomegranate fruit that has dark red, spherical seeds that resemble small garnet crystals.

Garnet has a long history of use in jewellery and as amulets throughout the world. Garnet is mentioned in the bible in chapter Exodus as part of the breastplate of Aaron. In the Koran they are the composition of the fourth heaven. In Vedic (Eastern) mythology hessonite garnet wards off evil.

There are also more exotic uses of synthetic garnets such as YIG (yttrium-iron-garnet). YIG has interesting magnetic properties and is used in microwave, magneto-optical, laser, and data storage applications. YAG, yttrium-aluminum-garnet, is used as synthetic gemstones that imitate diamonds, and in lasers as the optical medium.

KYANITE: Aluminum silicate: Al_2SiO_5

There is a series of minerals – sillimanite, andalusite, and kyanite – that all have the same chemical composition but different crystal structures; these are called polymorphs, meaning many forms. They form under differing metamorphic conditions. Of the three, sillimanite forms at the highest temperatures and pressures and andalusite at the lowest metamorphic intensity. Kyanite is the most common polymorph and it exhibits a unique physical property: variable hardness in different directions.

Appearance

Colour: Usually pale blue to distinct sapphire blue, also grey-white or very pale green.

Streak: White.

Lustre and transparency: Lustre is vitreous; pearly on cleavage faces. Transparent to translucent.

Habit: Often as long, bladed crystals. It may also be massive or fibrous.

Physical Properties

Hardness: Moderate to hard (Mohs 4 along cleavage planes, 6 across cleavage planes).

Density: Medium (3.6 g/cm^3).

Breakage: Brittle with one distinct cleavage parallel to elongation and an uneven fracture on the non-cleavage planes.

Test: Colour, as there are not many blue minerals. Variable hardness in different directions is important. It is found in metamorphic rocks with mica.

Kyanite (83909): Bara de Selinas, Coronel Murta, Minas Gerais, Brazil.
Width of field of view: 10.5 cm

Similar Minerals

Feldspar (p. 226) has two distinct cleavages, no differential hardness, and is only rarely pale blue.

Occurrence

Kyanite is found in metamorphic rocks such as mica schist (p. 286), and gneiss (p. 290).

Canada's Best Localities: Downie Creek intersecting Columbia River, Kootenay District, British Columbia; Canoe River near Valemont, Caribou District, British Columbia; Anderson Lake and Stall Lake Mines, near Snow Lake, Manitoba; Narco Mines, Témiscaming, Campeau Township, Témiscamingue County, Québec.

Other Localities: Judd's Bridge, Litchfield County, Connecticut, USA; Richmond County and Mitchell County, North Carolina, USA; Hampshire County, Massachusetts, USA; Minas

Gerais, Brazil; Tyrol, Austria; Pizzo Forno, Ticiano, Switzerland; Sultan Hamud, Kenya.

Kyanite (49775): Bladed crystal in mica schist. Anderson Lake Mine, Manitoba. Width of field of view: 5 cm

Interesting Facts

The mineral name is from the Greek, *kyanos*, meaning blue, which is the common colour for kyanite.

Kyanite is mined primarily for use in refractory materials that maintain their strength at high temperatures. Ceramic material such as dinnerware and plumbing fixtures may contain kyanite. Grinding mortars, electrical insulators, and spark plugs often have processed kyanite in them. Rarely, kyanite is cut as a gem primarily for gem collectors and not for jewellery, as the good cleavage makes it fragile.

Kyanite (42190): Blue prismatic crystals with good cleavage in quartz. Narco Mine, Campeau Tp., Témiscamingue Co., Québec. Width of field of view: 10 cm

TOPAZ: Aluminum silicate fluoride, hydroxide: $Al_2SiO_4(F,OH)_2$

Throughout history and right up to today, topaz has been noteworthy as a gemstone. It occurs in large, perfect crystals in a variety of colours. Some crystals from Brazil weigh over 200 kg. Topaz is often thought of as yellow in colour. Many misleading names such as smoky topaz and Madeira topaz were adopted to describe coloured quartz, smoky and citrine varieties respectively. Such usage of confusing mineral names is discouraged but they are given here as a buyer-beware note.

Topaz (58317): Ouro Preto, Minas Gerais, Brazil. Length of crystal: 9 cm

Appearance

Colour: Colourless, white to grey, blue to green, yellow to orange, and rarely pink.

Streak: White powder.

Lustre and transparency: Vitreous, and transparent to translucent.

Habit: Often as prismatic crystals that can be elongate or stubby. The prisms are terminated by pairs of faces that form a tent or wedge, or in some crystals a single planar face truncates the crystal.

Physical Properties

Hardness: Hard (Mohs 8).

Density: Medium (3.5 g/cm^3).

Breakage: Brittle with a perfect cleavage perpendicular to the prism length.

Test: Cleavage and hardness are important properties.

Similar Minerals

Quartz (p. 218) has no cleavage and it is less dense (2.65 g/cm^3). Quartz has hexagonal prisms while topaz has rhombic, four-sided prisms.

Occurrence

Topaz is found in igneous rocks with a lot of silica, i.e., rhyolite or granite pegmatite. It is often associated with feldspar (albite), mica, tourmaline, and cassiterite.

Topaz (53678): Gemmy rhombic crystal with pyramidal face.
Katalong, Mardan District, North-West Frontier, Pakistan.
Width of field of view: 3 cm

Topaz (45777): Vitreous lustre with pale blue tint. Prismatic crystal with tent (dome form) termination. Gem Claims, Seagull Batholith, Swift River, Yukon. Width of crystal: 2 cm

Canada's Best Locality: Crystals up to 5 cm have been found at Seagull Batholith, Yukon.
Other Localities: Thomas Range, Utah, USA; Minas Gerais, Brazil; Mursinka, Ural Mountains, Russia; Pakistan; Kabul Province, Afghanistan.

Interesting Facts

The mineral name topaz is from the Greek, *topazos*, meaning to seek. Topazos was the name of an island in the Red Sea that was apparently difficult to find, hence the necessity to "seek" for its location. This island is known today as Zaberget or Island of St. John and it is well known as a source of olivine (p. 164). In ancient times any yellow stone was known as topaz and undoubtedly this early reference was to a yellow-coloured olivine. Often topaz is irradiated in a nuclear generator to give it an intense, "electric" blue colour.

Topaz (45748): Rhombic prism with pyramidal termination. Note perfect cleavage plane parallel to the base of the crystal. Téofilo Otoni, Minas Gerais, Brazil. Width of crystal: 3 cm

TITANITE: Calcium titanium silicate: $CaTiSiO_5$

This relatively rare mineral has been found in exceptional localities in quantities large enough to be mined. Crystals can be fabulous in size and the lustre is unique. Titanite occurs in a wide variety of colours, the most attractive of which are green. In exceptional cases gems have been cut from transparent crystals. The bright crystals are a collector's item.

Appearance

Colour: Usually brown with a slight red tint but often almost black; sometimes green or yellow.

Streak: White.

Lustre and transparency: Adamantine lustre and transparent to translucent to almost opaque in deeply coloured specimens.

Habit: Often occurs as wedge-shaped crystals, sometimes twinned. Titanite can be compact and massive.

Physical Properties

Hardness: Hard (Mohs 5 – 5½).

Density: Medium (3.5 g/cm^3).

Breakage: Brittle with a distinct cleavage.

Test: A relatively high lustre for a transparent mineral.

Similar Minerals

Ilmenite (p. 78) resembles the darker colours of titanite but it has no distinct cleavage and it has a black streak. Ilmenite occurs in dark-coloured igneous rocks such as gabbro while titanite is in light-coloured igneous rocks.

Zircon (p. 178) has crystals that are square in cross-section and the cleavage is poor.

Magnetite (p. 74) is magnetic and has a more metallic lustre and a black streak.

Occurrence

Titanite is found in a wide variety of rocks, including igneous rocks such as granite and syenite and metamorphic rocks such as marble and schist.

Canada's Best Localities: Lake Harbour, Baffin Island, Nunavut; crystals up to 25 cm have been found at Tory Hill, Monmouth Township, Haliburton County, Ontario; Cardiff Uranium Mine, Cardiff Township, Haliburton County, Ontario; Lake Clear, Sebastopol Township, Renfrew County, Ontario; Yates Uranium Mine, Huddersfield Township, Pontiac County, Québec; Leslie Lake, Litchfield Township, Pontiac County, Québec.

Other Localities: Minas Gerais and Bahia, Brazil; Avenda, Aust-Agder, Norway; Salzburg, Austria; Val Tavetsch, Switzerland; Kola Peninsula, Russia.

Interesting Facts

The mineral is named for its titanium element content. Many textbooks still use the now obsolete name, sphene. The name "sphene" is from the Greek, *sphenos*, meaning wedge-shaped which describes the habit of the titanite crystals.

Like ilmenite (p. 78), and rutile (p. 118), titanite can be a source of the metal titanium. This lightweight, corrosion-resistant metal finds many uses in the aerospace industry, jewellery manufacture, and prosthetics. Because it is relatively soft and brittle, however, titanite does not wear well.

Among mineral collectors, crystalline specimens of titanite are highly prized. Some crystals can be cut into gemstones. Its optical properties are such that it has a greater dispersion than diamond. Dispersion is the ability of a material to divide or disperse white light into its coloured components, giving a gemstone "fire."

Titanite (84019): Wedge-shaped crystals with adamantine lustre on pale brown microcline. Bear Lake Rd., Gooderham – Tory Hill, Monmouth Tp., Haliburton Co., Ontario. Width of specimen: 20 cm

ZIRCON: Zirconium silicate: $ZrSiO_4$

Zircon is not a very common mineral but it is found in all of the different types of rock: igneous, sedimentary, and metamorphic. It always contains trace amounts of uranium and thorium, radioactive elements that may be used to date the rock or a geological event. It occurs as simple crystals with a beautiful lustre and reddish orange colour making them prized additions to a collection.

Appearance

Colour: Brownish-red, yellow-brown.

Streak: Too hard to streak but powder is white.

Lustre and transparency: Adamantine lustre on crystals. It is translucent to transparent.

Habit: Often as crystals: tetragonal prismatic with pyramid termination.

Physical Properties

Hardness: Hard (Mohs 7½).

Density: Medium (4.7 g/cm^3).

Breakage: Brittle with a poor cleavage on two planes.

Test: Colour, lustre, and hardness are important.

Zircon (37635): Kuehl Lake, Brudenell Tp., Renfrew Co., Ontario. Width of field of view: 8 cm

Similar Minerals

Titanite (p. 176) has wedge-shaped crystals and a perfect cleavage.

Rutile (p. 118) has a yellow streak and prismatic crystals are striated.

Ilmenite (p. 78) has a black to brownish streak and a conchoidal fracture.

Occurrence

It is found in many types of igneous rocks. Because of its hardness and durability it resists weathering sufficiently to be in sedimentary rocks.

Canada's Best Localities: Davis Quarry, Dungannon Township, Hastings County, Ontario; Yates Lake, Brudenell Township, Renfrew County, Ontario; McLaren Mine, Perth, Burgess Township, Lanark County, Ontario; Lake Clear, Sebastopol Township, Renfrew County, Ontario; Seybold Moore Mine, Val-des-Monts Township, Gatineau County, Québec; Lac Mathilda, Harrington Township, Argenteuil County, Québec; Mont-Saint-Hilaire, Rouville County, Québec.

Other Localities: Large perfect crystals from Seiland, Finnmark, Norway, and Betroka, Tuliara, Madagascar.

Interesting Facts

The mineral name is from Persian, *azargun*, *zar* meaning gold and *gun* meaning coloured.

Although the mineral was known in ancient times, its main constituent, zirconium, was not discovered as an element until 1789 by Martin Heinrich Klaproth.

Zirconium as an element does not occur in nature and its main source is the mineral zircon. Most of the zirconium metal is used to clad fuel rods in nuclear reactors since it is very transparent to neutrons that generate the heat in the reactor. It is used extensively in the chemical industry for pipes as it is very resistant to corrosion.

Zirconium dioxide, sometimes referred to as zirconia, has become one of the modern, wonder ceramics finding use as a thermal casing for jet engines, allowing them to operate at higher, hence more efficient, temperatures. As it has a high ionic conductivity it is useful as a electroceramic in oxygen sensors. It is also transparent to radio waves making it a durable replacement for plastic in iPod casings.

Zirconium dioxide can be grown as relatively large, cubic crystals at high temperatures. "Cubic zirconia" or "CZ" makes an effective substitute for diamond because of its high refractive index.

*Zircon (48361): Tetragonal prism and pyramid crystal with orange calcite.
Silver Queen Mine, Perth area,
North Burgess Tp., Lanark Co., Ontario.
Width of field of view: 7.5 cm*

EPIDOTE: Calcium iron aluminum silicate hydroxyl: $Ca_2(Al,Fe)_3(SiO_4)_3(OH)$

Epidote is a relatively common mineral in metamorphic rocks. It has a unique green colour which is likened to that of raw pistachio nuts. This pistachio green is sufficient to distinguish it from other green silicate minerals in the pyroxene and amphibole minerals groups.

Appearance

Colour: Yellowish green, dark green, greenish brown to almost black. The intensity of colour increases as the iron content increases.

Streak: Grey to white.

Lustre and transparency: Vitreous lustre and transparent to translucent.

Habit: Often as prismatic and tabular crystals. The crystals have a rectangular or square cross-section. Epidote may also be massive, granular, or fibrous in habit. Crystal faces are often striated parallel to the length.

Epidote (43605): Typical, pistachio green, granular epidote with pink calcite. Chelsea, Hull Tp., Gatineau Co., Québec. Width of field of view: 6.5 cm

Physical Properties

Hardness: Hard (Mohs 6–7).

Density: Medium (3.4 g/cm^3).

Breakage: Brittle with a distinct, single cleavage and an uneven or conchoidal fracture on the non-cleavage planes.

Test: Colour and hardness are important as well as the rock type in which it occurs.

Similar Minerals

Hornblende (p. 196) has two cleavages at 120º while epidote has only one cleavage. Hornblende may be magnetic as small splinters but epidote is never magnetic.

Diopside (p. 192) has two cleavages at 90º while epidote has only one cleavage.

Occurrence

Epidote is found in metamorphic rocks. In regionally metamorphosed rocks it is found with hornblende, albite, quartz, and chlorite. In marbles it may be found with garnet, vesuvianite, and diopside.

Canada's Best Localities: White River, Vancouver Island, British Columbia; Malone, Marmora Township, Hastings County, Ontario; Laugon Lake, Territoire-du-Nouveau Québec, Québec; Asbestos, Richmond County, Québec.

Other Localities: Excellent crystals from Prince of Wales Island, Alaska, USA; Calaveras County, California, USA; Mineral County, Nevada, USA; Mohawk, Idaho, USA; Baja

California, Mexico; Salzburg, Austria; Arendal, Norway; Le Bourg-d'Oisans, France; Piedmont, Italy; Windhoek, Namibia.

Interesting Facts

The mineral name is from the Greek, *epidosis*, meaning increase. The name refers to the fact that the rhombohedral crystal base has one pair of sides longer than the other.

Epidote is occasionally cut as gems.

Epidote (51629): Bladed crystals in rock fracture. Asbestos area, Shipton Tp., Richmond Co., Québec. Width of field of view: 13 cm

VESUVIANITE: Calcium magnesium iron aluminum silicate hydroxide: $Ca_{19}(Al,Mg,Fe)_{13}(SiO_4)_{10}(Si_2O_7)_4(OH)_{10}$

A somewhat rare, beautiful mineral, vesuvianite occurs in excellent crystals that are much sought by collectors. Vesuvianite is often called "idocrase,, but this is not the official IMA (International Mineral Association) name. The chemical composition is complex and the mineral has an unusual crystal structure that includes two types of silica groups: single SiO_4 tetrahedra and Si_2O_7 double tetrahedra. This structural feature is quite unique among minerals.

Appearance

Colour: Most often it is pale or dark green, but brown, yellow, white, and blue to purple are also found.

Streak: Vesuvianite is too hard to give a streak but the powder is white.

Lustre and transparency: Vitreous lustre, often with a somewhat greasy appearance, and transparent to translucent.

Habit: Often as prismatic crystals with a square cross-section and pyramidal terminations. It may also be massive.

Physical Properties

Hardness: Hard (Mohs 6).

Density: Medium (3.4 g/cm^3).

Breakage: Brittle with no distinct cleavage and an uneven or conchoidal fracture.

Test: Crystal form and hardness are important. The greasy lustre when visible is useful.

Similar Minerals

The pyroxene, diopside, (p. 192) often associated with vesuvianite, has two distinct cleavages that vesuvianite does not have; and the greasy lustre of vesuvianite is unlike the vitreous lustre of diopside.

Occurrence

Vesuvianite is found in contact metamorphic rocks of impure limestone.

Vesuvianite (83935): Long prism with pyramidal termination. Jeffrey Mine, Asbestos, Shipton Tp., Richmond Co., Ontario. Length of crystal: 6.5 cm

Canada's Best Localities: Silence Lake Mine, near Clearwater, Kamloops District, British Columbia; up to 11 cm crystals from W.F. Baxter Quarry, near Malone, Marmora Township, Hastings County, Ontario; Templeton, Hull

County, Québec; Jeffrey Mine, Asbestos, Richmond County, Québec, with transparent crystals of green, brown, and purple.

Other Localities: York County, Maine, USA; Essex County, New York, USA; Lamoille County, Vermont, USA; near Pulga, Butte County, California, USA; Lake Jaco, Chihuahua, Mexico; Piedmont, Italy; Yakutskaya, Russia; Balochistan, Pakistan.

Interesting Facts

The mineral name is from Mount Vesuvius, Italy, where it was found by Abraham Gottlob Werner in 1795. It occurs in blocks of metamorphosed limestone enclosed within the lavas.

Occasionally vesuvianite is faceted into gemstones and a compact, jade-like variety is sometimes carved. The variety name, "californite," has been applied to the jade-like material but usage of this term is discouraged.

Vesuvianite (55693): Jeffrey Mine, Asbestos, Shipton Tp., Richmond Co., Ontario. Width of field of view: 5.5 cm

BERYL: Beryllium aluminium silicate: $Be_3Al_2(SiO_3)_6$

Beryl is a mineral name unfamiliar to most people yet its variety names are common in the gem trade. Emerald is the variety name (i.e., not a species name) for dark green beryl; aquamarine is the variety name for blue-coloured beryl. Other varieties include heliodor (yellow, sometimes greenish-yellow beryl), morganite (pink beryl), and goshenite (colourless beryl). Pure beryl is colourless. The colours are due to small amounts of chemical impurities.

Beryl (81630): Etched, prismatic crystal Volodarsk-Volynskiy, Ukraine. Crystal length: 12 cm

Appearance

Colour: Colourless or white, red to pink, yellow or greenish yellow, pale green to deep (emerald) green and blue.

Streak: A streak is impossible due to the hardness but the powder of beryl is white.

Lustre and transparency: Vitreous lustre and transparent to translucent.

Habit: Often as crystals, commonly as hexagonal prisms terminated with a flat, pinacoid face or sometimes with a pyramid set of faces. The prism faces are often striated or grooved parallel to the long axis. Beryl can also be massive, making it more difficult to identify.

Physical Properties

Hardness: Hard (Mohs 7½ – 8).

Density: Medium (2.7 g/cm^3).

Breakage: Brittle with a conchoidal fracture.

Test: Hardness and crystal shape are important.

Similar Minerals

Quartz (p. 218) is quite similar to beryl but much more common. Quartz is often white or colourless while beryl is usually coloured bluish to greenish and is seldom colourless or white. Quartz usually has pyramidal terminations while beryl is often flat on the termination.

Apatite (p. 128) is softer (Mohs 5), and its crystals are not as sharp as beryl crystals. Apatite tends to have a slightly greasier appearance.

Tourmaline crystals (p. 188) are trigonal (3-sided) not hexagonal (6-sided) prisms.

Occurrence

Beryl is usually found in igneous, granitic rocks. It sometimes occurs in mica schist and the rare emeralds of Muso, Colombia, occur in a carbonaceous limestone.

Beryl (81633): Etched, stubby-prismatic crystal.
Chiá Mine, São José da Saira, Minas Gerais, Brazil.
Width of crystal: 5 cm

Canada's Best Localities: Quadeville, Lyndoch Township, Renfrew County, Ontario; Lacorne Township, Abitibi County, Québec.

Other Localities: Grafton, Cheshire and Sullivan Counties, New Hampshire, USA; Oxford County, Maine, USA; famous emeralds of Muso and Chivor, Colombia; Transvaal, South Africa; Bahia, Brazil; Jos, Nigeria; Sverdlovsk Oblast, Russia. Gemmy crystals have been found in Minas Gerais and Esperito Santo, Brazil; Nuristan, Afghanistan; Nagir and Peshwar, Pakistan; and the Ural Mountains, Russia. An 18 m long and 3.5 m wide crystal has been reported from Malakialina, Madagascar.

Interesting Facts

The name beryl is derived from Greek, *beryllus*, which describes the colour of the mineral as a precious blue-green colour of sea water.

Georgius Agricola (1546) noted an early use of a gemmy piece of beryl as a magnifying lens through which Nero was better able to see gladiatorial combats. This makes scientific sense as the higher refractive index of beryl would magnify an image. Other claims for the mineral may not be as easy to substantiate: improved intelligence, prevention of liver disease and epilepsy, improving one's manner, and protection against enemies.

Massive beryl is mined for the element beryllium. Because it is lightweight, it finds uses in alloys for missiles, high-speed aircraft, spacecraft, and communication satellites. It is also used in nuclear power plants as a neutron reflector. Beryllium copper alloy finds extensive uses in electrical contacts, springs, spot welding electrodes, and non-sparking tools.

As a gem, emerald can fetch prices beyond those of diamonds.

Beryl (37539): Flattened hexagonal prism of pink beryl (morganite variety). Tim Mine, Galiléia, Minas Gerais, Brazil. Width of field of view: 12 cm

Beryl (43101): Transparent, hexagonal prism with terminal pinacoid (emerald variety). Chivor Mines, Chivor, Boyaca, Colombia. Width of field of view: 2 cm

TOURMALINE GROUP

Sodium, calcium, aluminum, iron, magnesium, borate, silicate, hydroxide: $(Na,Ca)(Al,Fe,Mg,Li)_3(Al,Fe,Cr)_6(BO_3)_3Si_6O_{18}(OH)_4$

The tourmaline group of minerals is one of the most complex in mineral chemistry, having many different elements substituting in various atomic positions within the crystal structure. There are 12 species in the group. Tourmaline comes in a variety of beautiful colours and crystal forms making it an attractive mineral for collectors and a treasured gem in some cases.

Common tourmaline species include:

Schorl, $NaFe_3Al_6(BO_3)_3Si_6O_{18}(OH)_4$, black, most common tourmaline.

Dravite, $NaMg_3Al_6(BO_3)_3Si_6O_{18}(OH)_4$, brown.

Elbaite, $Na(Al,Li)_3Al_6(BO_3)_3Si_6O_{18}(OH)_4$, several shades of green and pink, sometimes cut as gems.

Appearance

Colour: Commonly black or brown, sometimes in shades of green, yellow, red, pink, and rarely blue. Some crystals are multicoloured along the length or across the length.

Streak: White.

Lustre and transparency: Vitreous lustre and transparent to translucent.

Habit: Often as elongate, prismatic crystals with pyramidal terminations. Crystals have a triangular cross section. Prism faces are striated parallel to the length of the crystal. On crystals terminated at both ends the faces do not match each other (termed hemimorphic or hemihedral, meaning each half of the crystal has its own, distinct symmetry).

Tourmaline (43172): Prismatic elbaite in granite. Lilypad Lakes, Kenora District, Ontario. Width of field of view: 11.5 cm

Tourmaline (37533): Multicoloured doubly terminated elbaite crystal with heavy striations on trigonal prism faces. Mesa Grande District, San Diego Co., California. Width of field of view: 8 cm

Tourmaline (50937): Prismatic dravite crystals with calcite. Tait Farm Prospect, Bancroft, Dungannon Tp., Hastings Co., Ontario. Width of field of view: 7.5 cm

Physical Properties

Hardness: Hard (Mohs 7 – 7½).
Density: Medium (3.1 g/cm³).
Breakage: Brittle with no distinct cleavage and a conchoidal or uneven fracture.
Test: Crystal form and hardness are important.

Similar Minerals

Vesuvianite (p. 182) crystals have a square cross-section white tourmaline is triangular. Pyroxene (p. 192) and amphibole (p. 196) have good cleavage while tourmaline does not.

Occurrence

Tourmaline is found in igneous rocks, most commonly granite. It is also sometimes found in metamorphic rocks such as marble or gneiss.

Canada's Best Localities: Schorl from Villeneuve Mine, Papineau County, Québec. Dravite from Tait Farm, Dungannon Township, Hastings County, Ontario; Enterprise, Sheffield Township, Lennox and Addington County, Ontario. Elbaite from Leduc Mine, Wakefield Township, Gatineau County, Québec; O'Grady Batholith, Northwest Territories.

Other Localities: Schorl from Mount Mica, Oxford County, Maine, USA; Riverside County, California, USA. Dravite from Pierrepont, St. Lawrence County, New York, USA; Yinnietharra, Western Australia, Australia; Rio Grande do Norte, Brazil. Elbaite from San Diego County, California, USA; Cumberland County and Oxford County, Maine, USA; Minas Gerais, Brazil; Laghman, Afghanistan; Madagascar; Sri Lanka; Elba, Italy; Nampula, Mozambique.

Interesting Facts

The name tourmaline is derived from the Sinhalese word, *turamali*, meaning stone attracting ash in reference to its pyroelectric properties.

Hemimorphic crystals like tourmaline have interesting electrical properties. They are both piezoelectric, forming an electric charge when subjected to pressure, and pyroelectric, producing an electric charge when heated. In some museum displays the heat from lighting is sufficient to induce pyroelectric effects, attracting dust particles to one end of a crystal.

Gem varieties of tourmaline are very popular due to their variety of colours and relatively low cost. There are several variety names such as "rubellite" for pink or rose elbaite, "indicolite" for dark blue or indigo elbaite, and watermelon tourmaline which is bicoloured elbaite, green on the outside with a pink interior.

Tourmaline (85171): Dravite Poudrette Quarry, Mt. St.-Hilaire, Rouville Co., Québec. Width of field of view: 24 cm

PYROXENE GROUP

Calcium, sodium, magnesium, iron, aluminum, silicate: $(Ca,Na)(Fe,Mg)(Si,Al)_2O_6$

Pyroxene is an important and common group of minerals with 22 species in the group. They are commonly found in many igneous and metamorphic rocks. Their presence indicates that the rock formed at temperatures high enough to drive off all water. If water were present, members of the amphibole group (p. 196) would crystallize. To determine the conditions under which a rock formed, it is important to be able to distinguish pyroxene from amphibole. Ore deposits like those at Sudbury, Ontario, that are rich in nickel, copper and platinum metals, are associated with gabbro (p. 252), a rock rich in pyroxene and poor in amphibole.

Common mineral species in the pyroxene group:

Diopside $CaMgSi_2O_6$; white to green; metamorphosed limestones (marble).

Augite $(Ca,Na)(Fe,Mg,Al)(Si,Al)_2O_6$; dark green to almost black; igneous rocks such as gabbro, pyroxenite, and basalt.

Aegirine $NaFeSi_2O_6$; dark green to almost black; in sodium-rich igneous rock such as syenite.

Jadeite $Na(Al,Fe)Si_2O_6$; white or light to dark green; rare mineral in high-pressure metamorphic rocks.

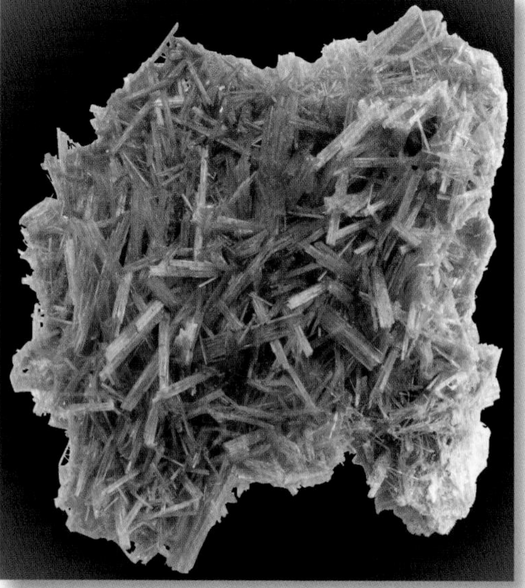

Pyroxene (32815): Needle-shaped diopside crystals. Jeffrey Mine, Asbestos, Shipton Tp., Richmond Co., Québec. Width of field of view: 8.5 cm

Appearance

Colour: White, pale to dark green, brown to black.

Streak: Greyish white to pale greenish white.

Lustre and transparency: Vitreous lustre and transparent to translucent.

Habit: Often as prismatic or stubby crystals. It is commonly massive in compact and granular varieties.

Physical Properties

Hardness: Moderate hardness (Mohs 5 – 6).

Density: Medium (3.3 – 3.5 g/cm³).

Breakage: Brittle with two distinct cleavages meeting at roughly 90º.

Test: Cleavage is important.

Similar Minerals

Amphibole (p. 196) has two cleavages at 60° or 120º to each other.

Epidote (p. 180) has only one cleavage and it tends to be more yellow-green than pyroxene.

Pyroxene (51677): Diopside crystals with pink calcite. Prismatic with a pyramidal terminati
Note two good cleavages at 90° in broken portion of specimen.
Laurel-Montfort Rd., Laurel, Wentworth Tp., Argenteuil Co., Québec.
Width of field of view: 5 cm

Pyroxene (41510): Dark green, prismatic augite crystals. Lake Clear, Sebastopol Tp., Renfrew Co., Ontario. Width of specimen: 10 cm

Occurrence

Pyroxene is found in a wide variety of igneous and metamorphic rocks. In igneous rocks such as gabbro (p. 252), peridotite, and basalt (p. 262), pyroxene is a common constituent. In metamorphic rocks pyroxene is often a constituent in high temperature-altered limestone, or marble (contact metamorphism).

Canada's Best Localities: Augite from Smart property, Sebastopol Township, Renfrew County, Ontario. Diopside from Soper River, Baffin Island, Nunavut; Birds Creek, Herschel Township, Hastings County, Ontario; Wilberforce, Dudley Township, Haliburton County, Ontario; Yates Uranium Mine, Huddersfield Tp., Pontiac County, Québec; Laurel, Wentworth Township, Argenteuil County, Québec; Lac Girard Mine, Wakefield Township, Gatineau County, Québec; Cawood, Cawood Township, Pontiac County, Québec; Orford Nickel Mine, Orford Township, Sherbrooke County, Québec. Aegirine from Mont Saint-Hilaire, Rouville County, Québec. Jadeite from Mount Ogden, Cassiar District, British Columbia.

Other Localities: Augite from Salzburg, Austria; Gilgit, Pakistan. Diopside from Dekalb Township, St. Lawrence County, New York, USA; Tyrol, Austria; Piedmont, Italy; Yakutskaya, Russia; Jaipur, India. Aegirine from Narssârssuk, Greenland; Langesundfjord, Norway. Jadeite from Burma.

Interesting Facts

The mineral name pyroxene is from the Greek, *pyr*, meaning fire and *xenos*, meaning stranger. Small crystals of the mineral were found in volcanic glass (hence fire) and were thought to be inclusions and not really belonging in the volcanic glass. Now we know that the pyroxene is one of the first minerals to crystallize in a magma and when the magma erupts it may be the only crystal formed, the rest solidifying as glass.

The pyroxene species, jadeite, is very rare. It is greatly valued as gem jade or imperial jade as it is called to distinguish it from the more common nephrite jade (p. 198). The history of jade dates back to 500 BCE in China, and there are numerous Chinese jade carvings. Both types of jade occur in China so it is not always possible to distinguish which type is being used.

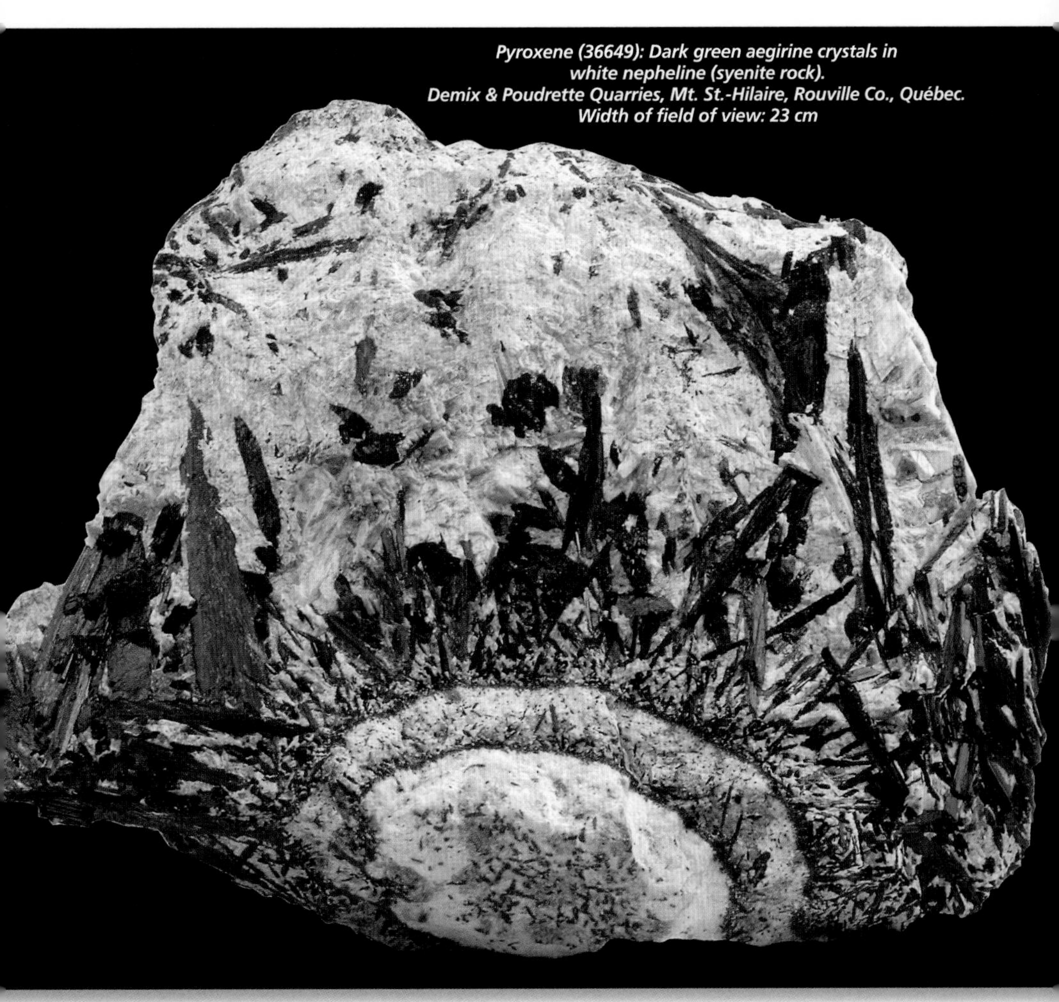

Pyroxene (36649): Dark green aegirine crystals in white nepheline (syenite rock).
Demix & Poudrette Quarries, Mt. St.-Hilaire, Rouville Co., Québec.
Width of field of view: 23 cm

AMPHIBOLE GROUP

Calcium sodium iron magnesium aluminium silicate hydroxide: $(Ca,Na)_2(Fe,Mg)_5(Si,Al)_8O_{22}(OH)_2$

Amphibole is a large mineral group of some 94 species. In general they are dark in colour, complex in chemical constituents, and resemble the chemically similar pyroxene group of minerals. Both groups are found in a wide variety of rocks; each indicates a different environment of formation.

Amphiboles crystallize from fluids that have more water and are lower in temperature than those solutions required for pyroxene crystallization. This increase in water content is important in the formation of certain ore deposits. The huge porphyry copper deposits of the southern USA are associated with diorite (p. 250) rocks that have dominant amphibole and little or no pyroxene for the dark-coloured minerals.

Some of the common amphibole species are hornblende, tremolite, actinolite, anthophyllite, riebeckite, arfvedsonite, richterite, and edenite.

Amphibole, edenite (51157): Prismatic crystals on pink calcite. George Earle Farm Property, Wilberforce, Monmouth Tp., Haliburton Co., Ontario. Width of field of view: 11 cm

Edenite (38647): Prismatic crystal.
Kuehl Lake, Brudenell Tp., Renfrew Co., Ontario.
Width of crystal: 9 cm

Actinolite, nephrite jade variety (45207): Water worn jade Fraser River, near Hope, Yale District, British Columbia. Width of specimen: 14 cm

Appearance

Colour: Pale green to dark green, almost black.

Streak: Greyish white, sometimes with a hint of green if the specimen is dark coloured.

Lustre and transparency: Vitreous lustre and translucent to opaque.

Habit: Often as blocky crystals (hornblende, richterite) or prismatic crystals (tremolite, actinolite). It may also be massive and fibrous (riebeckite, tremolite, anthophyllite). When the fibres become interlocked, the result is a very compact variety of tremolite known as jade or nephrite.

Physical Properties

Hardness: Moderate (Mohs 5½ – 6).

Density: Medium (2.9 – 3.8 g/cm^3). The large variation in density is due to the great number of chemical substitutions that are possible in this group.

Breakage: Brittle with two good cleavages and a conchoidal fracture on the non-cleavage planes. The elongate cleavage fragments have a rhombic cross-section with angles measuring 120° or 60°.

Amphibole, riebeckite, crocidolite variety (oj1243): Fibrous bands of the blue variety of asbestos, crocidolite. Griqualand, Republic of South Africa. Width of field of view: 4 cm

Test: Streak and cleavage are the most important criteria. Amphiboles such as hornblende and tremolite are rich in iron (Fe) and small fragments may be slightly magnetic.

Similar Minerals

Pyroxene group minerals (p. 192) have two cleavages that intersect in cross-section at 90° angles.

Fibrous riebeckite resembles fibrous chrysotile (p. 214) but the fibres are much stiffer than those of chrysotile.

Epidote (p. 180) has only one cleavage while amphibole has two cleavages.

Amphibole, edenite (38647): Cleavage of 120° in the top half of crystal. Kuehl Lake, Brudenell Tp., Renfrew Co., Ontario. Width of crystal: 9 cm

Occurrence

It is found in all sorts of igneous rocks such as granite, basalt, and syenite; also in metamorphic rocks such as marble, schist, gneiss, and amphibolite.

Canada's Best Localities: Actinolite, variety nephrite jade, is mined at Mount Ogden, Cassiar District, British Columbia; sharp tremolite and actinolite crystals at Bob's Lake and Sharbot Lake, Oso Township, Frontenac County, Ontario; large hornblende crystals over 15 cm have been found at Tory Hill, Monmouth Township, Haliburton County, Ontario; long fibres of tremolite in the Jeffrey Mine, Shipton Township, Richmond County, Québec.

Other Localities: Purple crystals of tremolite at Balmat, Fowler Township, St. Lawrence County, New York, USA; blue asbestos at Penge, Transvaal, South Africa, and Wittenoon Gorge, Western Australia; pink tremolite crystals from Kunar, Afghanistan; actinolite jade from Irkutskaya Oblast', Russia.

Interesting Facts

The mineral name is from the Greek *amfibolos*, meaning ambiguous, as it may be confused with other minerals.

The compact variety of nephrite jade was used in tools, carvings, and jewellery by the Native People of the Canadian Arctic and the Northwest Coast, and by the Maori of New Zealand. Nephrite jade, an amphibole, should not be confused with the rarer and more expensive jadeite (a pyroxene) that is often called Imperial jade and comes almost exclusively from Burma.

RHODONITE: Manganese silicate: $MnSiO_3$

Rhodonite is a relatively rare mineral but it is often seen in carvings and jewellery. It is pink like the other manganese mineral rhodochrosite, with which it is often confused. Crystals are very rare and much sought after for collections.

Appearance

Colour: Most commonly pink but it may be in shades of rose-red or red-brown. Often rhodonite has inclusions of manganese oxide that show up as black patches or lines.
Streak: White.
Lustre and transparency: Vitreous lustre and translucent.
Habit: Often massive and fine-grained. Crystals are rare with tabular or blocky prismatic form.

Physical Properties

Hardness: Hard (Mohs 6).
Density: Medium (3.6 g/cm^3).
Breakage: Crystals are brittle with two distinct cleavages at right angles. Massive forms have a conchoidal fracture.
Test: Colour and hardness are important. The black alteration lines are characteristic.

Similar Minerals

Rhodochrosite (p. 152) is softer (Mohs 3 – 4) has three cleavages that meet at 120º angles forming a rhombohedron. Rhodochrosite does not have the black inclusions that rhodonite often does and it fizzes in acid while rhodonite does not fizz.

Occurrence

Rhodonite is found in metamorphic rocks that are rich in manganese.

Canada's Best Localities: Saltspring Island, British Columbia; Keremeos, Similkameen Division, Yale District, British Columbia; Arthur Point, Coast District, British Columbia; Williams Lake, Cariboo District, British Columbia; Lac Jeannine, Conan Township, Saguenay County, Québec.
Other Localities: Franklin, New Jersey, USA; Plainfield, Hampshire County, Massachusetts, USA; Broken Hill, New South Wales, Australia (Australia has large crystals); Ural Mountains, Russia; Kaso Mine, Kanuma, Japan.

Interesting Facts

The mineral name is from the Greek, *rhodon*, meaning rose.

Due to the colour, hardness, and fine-grained massive nature of rhodonite, it is used in carvings and beads.

Rhodonite (44197):
Saltspring Island, British Columbia.
Width of field of view: 16 cm

MICA GROUP

Potassium, iron, magnesium, aluminum, silicate, hydroxide:
$K(Fe, Mg, Al)_2[(Si,Al)_4O_{10}](OH)_2$

Mica is a large group of approximately 30 mineral species. They are important in all types of rocks – igneous, metamorphic, and sedimentary. All micas have a layered crystal structure giving them a perfect cleavage that gives distinctive, very thin sheets.

There are three common micas:

Muscovite: $KAl_2[AlSi_3O_{10}](OH)_2$; usually colourless or white to silvery.

Biotite: $K(Fe, Mg)_3[AlSi_3O_{10}](OH)_2$; black to dark brown colour.

Phlogopite: $K(Mg, Fe)_3[AlSi_3O_{10}](OH)_2$; brown to tan in colour.

Appearance

Colour: Colour can be used as a preliminary guide to which species of mica you have. It may be colourless, white, silvery, yellow, brown, or black. More exotic species of mica can be mauve (named lepidolite) or green (named celadonite).

Streak: White powder. Difficult to grind by hand or mechanically due to perfect cleavage.

Lustre and transparency: Vitreous to pearly lustre and transparent to translucent.

Habit: Often occurring as diamond- and hexagonal-shaped, prismatic crystals with a flat termination (pinacoid). Prism faces are not bright but are undulating and dull due to the dominant cleavage. Also found as laminated masses.

Physical Properties

Hardness: Moderate (Mohs 2 – 3).

Density: Medium (2.8 g/cm^3 average).

Breakage: Easily separated with a distinct cleavage. Cleavage folia are flexible and elastic.

Test: Cleavage and elasticity of cleavage plates are important.

Similar Minerals

The mica species can be differentiated by colour and mode of occurrence. Weathered biotite sometimes appears shiny yellow, misleading people to think they have found gold.

Chlorite (p. 208) has a perfect cleavage but the cleavage fragments are not elastic like those of mica.

Occurrence

Mica is found in igneous, metamorphic, and sedimentary rocks yet some environments help distinguish which mica species you have. Muscovite is commonly found in granite or related pegmatites and in metamorphic schists. Biotite is also found in granite but it can be in more silica-poor rocks such as gabbro and in metamorphic schists. Phlogopite is found in very silica-poor rocks such as peridotite and pyroxenite.

Canada's Best Localities: Muscovite: Mica Mountains, Cariboo District, British Columbia; Purdy Mine, Mattawa Township, "ruby mica" from McAuslan Township, and Maskwa Lake

Mica, biotite (50499): Hexagonal, prismatic crystal with platy steps. Giroux Mine, Otter Lake area, Pontiac Co., Québec. Width of specimen: 7 cm

Mine, Deacon Township, Nipissing District, Ontario; Craigmont Corundum Mine, Raglan Township, Renfrew County, Ontario; Kasshabog Lake, Methuen Township, Peterborough County, Ontario; Villeneuve Mine, Villeneuve Township, Papineau County, Québec. Phlogopite: Kimmirut, Baffin Island, Nunavut; March Township, Carleton County, Ontario; Grenville Township, Argenteuil County, Québec, Blackburn Mine, Hull Township, Gatineau County, Québec. Biotite: Davis Hill, Dungannon Township, Hastings County, Ontario.

Other Localities: Muscovite: Montgomery County, Pennsylvania, USA; Magnet Cove, Arkansas, USA; Grafton County, New Hampshire, USA; Lincoln County, North Carolina, USA; Governador Valadares, Minas Gerais, Brazil; Madras, Tamil Nadu, India. Phlogopite: Andranondambo, Madagascar. Biotite: With muscovite Thomaston, Upson County, Georgia, USA.

Interesting Facts

The word mica is from the Latin, *micare*, meaning to shine. The name muscovite is from Muscovy glass, as it was used for window panes in Russia.

Mica is a good insulator of heat and electricity. As a result, it was used in the past in stove windows, toasters, high-voltage insulators, and radio capacitors.

Mica was also used in ancient times. Cave artists of the Upper Paleolithic period (40,000 to 10,000 BCE) used mica to produce white pigment. Just north of Mexico City stands the Pyramid of the Sun, which belonged to Teotihuacan culture. The entire top level of the pyramid is a 30 cm thick slab of mica. Near Trivandrum, India, is the Padmanabhapuram palace whose coloured windows are mica.

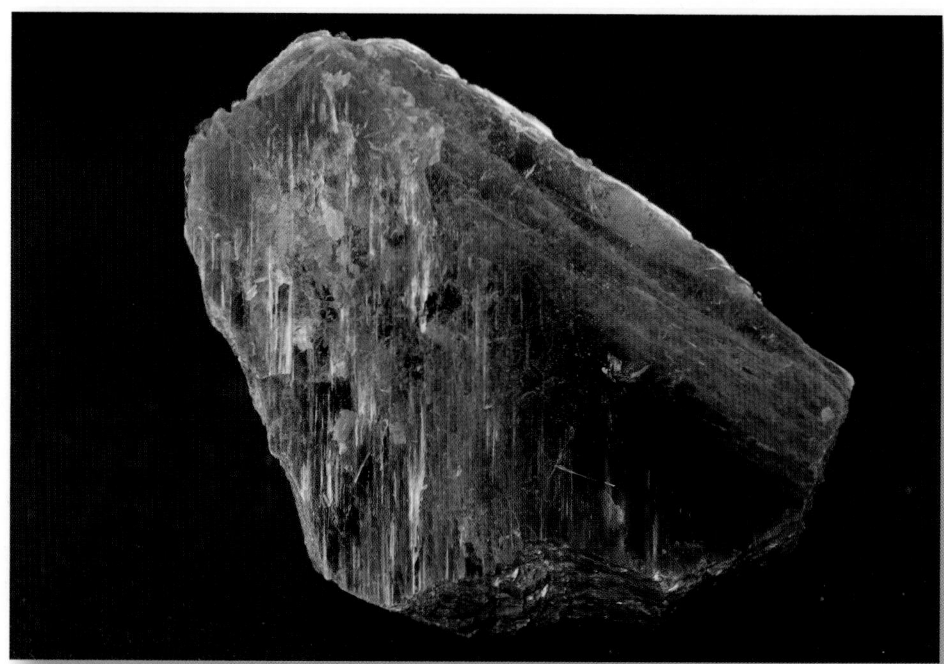

Mica, muscovite cleavage plate showing a twin (50070): Wanup, Dill Twp., Sudbury District, Ontario. Width of field of view: 12 cm Photo: R.A. Gault

Mica, phlogopite (46668): Platy, hexagonal crystal with perfect cleavage. Soper River, N. of Kimmirut, Baffin Island, Nunavut. Width of specimen: 4 cm

CHLORITE GROUP

Iron, magnesium, aluminum, silicate, hydroxide:
$(Fe,Mg,Al)_6(Si,Al)_4O_{10}(OH)_8$

Chlorite is a group of minerals consisting of 12 species. As a group they all have the same layered crystal structure and the same platy morphology, but individual species within the group are difficult to distinguish. Some rarer species of chlorite have significant amounts of zinc, manganese, lithium, and nickel in the crystal structure. Chlorite is indicative of rock altered by hydrothermal solutions.

Appearance

Colour: As the mineral increases in iron content (decreases in magnesium) the colour changes from white or yellowish to pale green, green to almost black.

Streak: White or pale green.

Lustre and transparency: Vitreous lustre for coarse plates and earthy for fine-grained varieties and transparent to almost opaque.

Habit: Often as platy, hexagonal crystals. It may also be massive and earthy.

Physical Properties

Hardness: Moderate (Mohs 2 – 2½).

Density: Medium (2.8 g/cm³).

Breakage: A distinct cleavage. Plates are flexible but not elastic.

Test: Cleavage and flexible (but not elastic) plates are important properties.

Similar Minerals

Mica (p. 204) has a perfect cleavage and its cleavage fragments are flexible and elastic. The cleavage fragments of chlorite are not elastic.

Talc (p. 210) has a cleavage like that of chlorite but it is softer (Mohs 1) and feels greasy. The two may be intergrown.

Occurrence

Chlorite is commonly found in metamorphosed rocks that have undergone relatively low temperature and pressure effects. It occurs with quartz, albite feldspar, and garnet in schists. Chlorite is also found in igneous rocks as an alteration product of minerals such as pyroxene, amphibole, and biotite mica. It is a common alteration mineral associated with hydrothermal ore deposits and commonly occurs with epidote. It is also an important constituent of sediments.

Canada's Best Localities: Blackstone Lake, Conger Township, Parry Sound District, Ontario; Broughton Soapstone Quarry, Leeds Township, Mégantic County, Québec.

Other Localities: Bristol County, Massachusetts, USA; San Benito County, California, USA; Lancaster County and Chester County, Pennsylvania, USA; purple variety from Erzurum, Eastern Anatolia, Turkey.

Interesting Facts

The mineral group name chlorite comes from Greek, *chloros*, meaning green, the usual colour of this group of minerals.

Chlorite inclusions in clear quartz are particularly interesting as they may produce a "phantom" when a crystal appears to include another crystal.

Kammererite is a variety in the chlorite group that contains chromium in sufficient amounts to colour specimens a deep purple. These are sought by collectors.

Chlorite (47089): Pearly lustre, perfect cleavage.
York River, Bancroft, Ontario. Width of specimen: 4 cm

TALC: Magnesium silicate hydroxide: $Mg_3Si_4O_{10}(OH)_2$

This mineral is known to most of us for its use in talcum powder or baby powder. One of the softest minerals, it has become the test mineral for Mohs hardness of 1. Talc is almost never found as crystals but as a fine-grained compact aggregate. It is always found in metamorphic rocks as an alteration product.

Appearance

Colour: Usually pale green but also grey or white.

Streak: White.

Lustre and transparency: Waxy, pearly, or greasy lustre and translucent.

Habit: Often massive with either a compact or foliated (layered) habit. Rarely found as tabular crystals.

Physical Properties

Hardness: Soft (Mohs 1) and feels greasy or soapy. Massive varieties are easily cut with a knife (sectile).

Density: Medium (2.7 g/cm^3).

Breakage: Flexible but inelastic sheets or folia separate easily with a perfect cleavage.

Test: Hardness and greasy feel.

Similar Minerals

Muscovite mica (p. 204) is harder (Mohs 2), and has elastic folia or cleavage flakes. Fine-grained muscovite does not feel greasy like talc does.

Clay (p. 212) tends to be white, not green like talc. When the powder is wet, clay behaves differently from talc, in that it can be moulded and talc cannot.

Chlorite (p. 208) is harder (Mohs 2 – 2½)

Occurrence

Talc is found as the alteration (metamorphic) product of rocks such as pyroxenite and peridotite. This alteration takes place through the action of hydrothermal solutions.

Canada's Best Localities: Cassiar Mine, Cassiar District, British Columbia; Grimsthorpe Township, Huntington Township, and Elzevir Township, Hastings County, Ontario.

Other Localities: Belmont, St. Lawrence County, New York, USA; Washington County and Windsor County, Vermont, USA; Providence County, Rhode Island, USA; Sel Herad, Oppland, Norway; Zillertal, Tyrol, Austria.

Interesting Facts

The mineral name is from the Arabic, *talq*, meaning pure, and perhaps referring to the white colour of its powder.

Because of its unique properties, talc is widely used as filler in paper, paint, pharmaceuticals, cosmetics, and animal feed. As it does not conduct heat or electricity, it is used in stoves and electrical panels. A common variety used in many types of carving is soapstone. Sometimes soapstone is called steatite, but this name could refer to talc or to harder varieties containing chlorite or serpentine minerals.

There has been concern in recent years about the use of talcum powder since in many cases it may be intimately associated with asbestos-like minerals, which may be carcinogenic.

Talc (45763): Flexible sheet with pearly lustre. Hastings Co., Ontario. Width of field of view: 27 cm

CLAY GROUP

Aluminum, silicate, hydroxide or hydrate: $Al_2Si_2O_5(OH)_4 \cdot H_2O$

Clay is a general term applied to an important group of mineral species. Clays are well known for their widespread occurrence in soils and rocks and their vast history in pottery and ceramics. Like the mica and chlorite groups, the clay minerals have a layered crystal structure within which the layers are weakly bonded together, giving it special, useful properties. Clays are very similar to mica except they have the ability to absorb and retain water. The clay minerals are not spectacular in a collection but they are of such importance in their applications that one should learn to identify them.

The three most common clay mineral species are:
Kaolinite: $Al_2Si_2O_5(OH)_4$
Montmorillonite: $NaAlSi_2O_5(OH) \cdot H_2O$
Illite: $KAl_2(Si_3Al)O_{10}(H_2O,OH)_2$

Appearance

Colour: Pure clay is white but it may be tinted reddish or brownish with impurities such as iron.
Streak: White.

Kaolinite clay: Feldspar altered to kaolinite leaving quartz grains unaltered. St. Stephens, Cornwall, England. Width of specimen: 8 cm

Lustre and transparency: Earthy and dull when fine-grained but may have a pearly lustre in coarser grains. It is translucent.
Habit: Often massive and very fine-grained.

Physical Properties

Hardness: Hardness is almost impossible to discern as it is usually so fine-grained that scratching merely separates grains. Soft (Mohs about 2).
Density: Medium (2.6 g/cm^3).
Breakage: Chalky and easily broken when dry but this changes when wet, becoming elastic.
Test: Slippery and flexible when wet.

Similar Minerals

Mica (p. 204) tends to have bigger flakes that are flexible and elastic.
Talc (p. 210) is usually green in colour and coarser than clay particles.
Neither mica nor talc has the sticky or cohesive property of wet, powdered clay.

Occurrence

Clay minerals are found in weathered or hydrothermally altered rocks that were originally rich in aluminum and silicon.

Canada's Best Localities: Kaolinite occurs at Walton, Hants County, Nova Scotia and at Bonavista Bay, Newfoundland.
Other Localities: Kaolinite deposits at Twiggs County, Georgia, Lawrence County, Indiana, and Talladega County, Alabama, USA. A classic locality for kaolinite replacing orthoclase feldspar is at St. Austell, England.

Interesting Facts

The word clay is from the old English, *claeg* meaning stiff or sticky earth. This would certainly describe one of the main properties of this mineral group, a property that was known from prehistoric times and confirmed with the excavation of sun-dried bricks and drinking vessels from early settlement sites.

The term China clay (kaolinite) comes from the use of clay in 7th- and 8th-century Chinese porcelains. China clay was mined in Jiangxi Province in southeastern China. Bone china was developed early in the 1800s by Josiah Spode II. He found that adding about 50 percent bone material to kaolinite clay increased its translucency and strength.

Today clay is still utilized in ceramic and brick materials. Other uses include fillers in paper, paint, plastic, and rubber production. Certain clays have the ability to exchange cations in their structure. Montmorillonite and illite can exchange sodium and potassium for other, more highly-charged ions. The resulting clays are used to absorb radioactive elements in nuclear waste spills or oil spills.

CHRYSOTILE: Magnesium hydroxyl silicate: $Mg_3Si_2O_5(OH)_4$

Chrysotile (36490): Golden, fibrous habit. Bowman Mine, Timmins area, Cochrane District, Ontario. Width of specimen: 16 cm

Chrysotile is a member of the serpentine group. The fibrous variety is also known as asbestos, and most of the asbestos mined today is chrysotile. However, the term asbestos can also refer to several different mineral species, most of which are in the amphibole group. Unfortunately the physical properties that make asbestos useful also make it a carcinogen. Fibres capable of piercing lung tissue promote cancerous growth. Chrysotile is the least harmful type of asbestos.

Appearance

Colour: Usually green but it may be white, grey, golden yellow, or brown.

Streak: White.

Lustre and transparency: The lustre is silky when fibrous and greasy when massive and translucent.

Habit: Often fibrous but it may be massive.

Physical Properties

Hardness: Moderate (Mohs 2 – 3).

Density: Medium (2.6 g/cm^3).

Breakage: No distinct cleavage separates into fibres.

Test: Fibres are soft and flexible.

Similar Minerals

Riebeckite (p. 196), an amphibole, can be fibrous but the fibres are stiffer and less flexible.

Occurrence

Chrysotile is found in ultrabasic rocks altered by hydrothermal solutions.

Canada's Best Localities: Chrysotile is mined at the Jeffrey Mine, Asbestos, Shipton Township, Richmond County, and Thetford, Mégantic County, in southern Québec, and at Cassiar in northern British Columbia. Near Timmins, Deloro Township, Cochrane District, Ontario; Munro Mine, Cochrane District, Ontario; Kilmar, Grenville County, Ontario; Argenteuil County, Québec.

Other Localities: Globe, Gila County, Arizona, USA; Piedmont and Lombardia, Italy; Shabani, Zimbabwe.

Interesting Facts

The mineral name is from the Greek, *chrysos*, meaning gold, plus, *tilos*, meaning fibre, referring to its colour and habit although it is seldom golden-coloured. Asbestos derives from the Greek word, *asbestos*, meaning incombustible. The description comes from Pliny (CE 23 to CE 73), who refers to a rare cloth, *linum vivum* (immortal linen), which the Romans wove into a cremation shroud for their kings. During cremation, the chrysotile shroud would survive intact and the body would turn to ash. Today, asbestos is used in water filters, brake pads, heat insulation in home electrical appliances and rockets, reinforcement in cement blocks and inflammable materials.

Chrysotile (44536): Ribbons of fibrous type in a more compact variety. Normandie Mine, Black Lake, Ireland Tp., Mégantic Co., Québec. Width of field of view: 19 cm

CHRYSOCOLLA: Copper-hydrate silicate: $(Cu,Al)_2H_2Si_2O_5(OH)_4.nH_2O$

Chrysocolla is an attractive blue-green mineral that is distinctive in colour and form. It is not well understood as a mineral as it is not well crystallized and is often amorphous. For many years it has been used by prospectors as an "indicator" mineral for copper. Many specimens are covered in quartz, giving rise to false hardness readings.

Appearance

Colour: Green to blue-green. It may also be black to brown if it has many inclusions.
Streak: White to blue-green.
Lustre and transparency: Vitreous to earthy to waxy lustre and transparent to opaque.
Habit: Often massive but also found as botryoidal crusts or seam fillings. More rarely it may be found as fine fibres.

Physical Properties

Hardness: Soft to medium (Mohs 2 – 4).
Density: Light (2.0 g/cm^3).
Breakage: Brittle with a conchoidal fracture and no cleavage.
Test: Colour, softness, lack of crystals, and fragility are good field indicators for this mineral.

Similar Minerals

Turquoise (p. 136) is harder (Mohs 5 – 6) and usually different in colour.
Chrysoprase is a variety of fine-grained quartz that is similar in colour but much harder (Mohs 7).

Chrysocolla (56136): Botryoidal crust.
Likasi Mine, Likasi, Democratic Republic of Congo. Width of field of view: 12 cm

Occurrence

Chrysocolla is found in altered copper deposits often associated with quartz, azurite, malachite, and limonite.

Canada's Best Localities: Craigmont Copper Mine, Kamloops District, British Columbia; Gaspé Copper Mine, Holland Township, Gaspé-Ouest County, Québec.

Other Localities: Arizona, Utah, and New Mexico, USA, all have considerable amounts of chrysocolla associated with copper deposits; San Carlos, Mexico; Kakanda, Katanga Province, Democratic Republic of Congo.

Interesting Facts

The origin of the name chrysocolla is rather strange. It was derived during Greek antiquity from the Greek, *chrysos*, gold, and *kolla*, glue, as a result of its use along with several green minerals containing copper as a glue, or flux, for soldering gold.

Pure chrysocolla is too soft for use in jewellery. However, chrysocolla is often intermixed with fine-grained quartz (variety name chalcedony) making it harder and durable enough to be polished. Occasionally, chrysocolla is turquoise in colour and is used as a substitute for gemmy turquoise which is much more valuable.

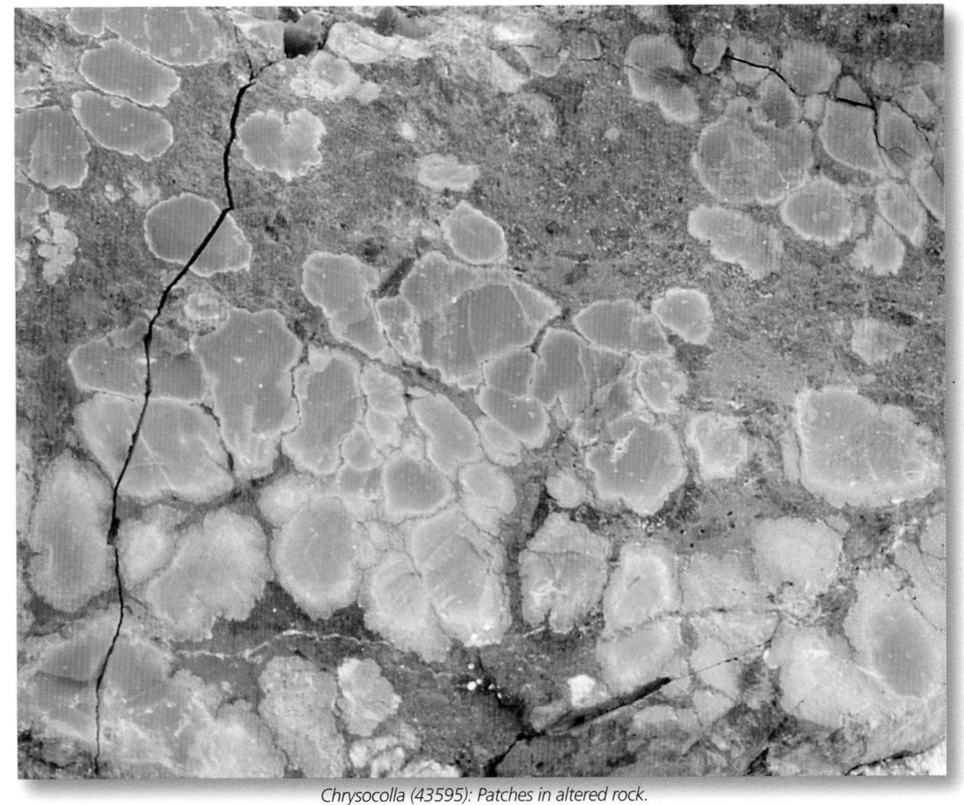

Chrysocolla (43595): Patches in altered rock.
Lornex Mine, Highland Valley, Kamloops Division. Yale District, British Columbia. Width of field of view: 4 cm

QUARTZ: Silicon oxide: SiO_2

Quartz is one of the most common minerals, being surpassed only by the feldspar group for its occurrence on the earth's surface. It occurs in such a wide variety of forms and colours that it can be a challenge to identify. Many mineral collectors have large specialized collections of quartz as it is fun to compare the innumerable possibilities of just one species. It is inexpensive and can be found everywhere.

There are two broad categories in the quartz family based on crystal size – large crystal (macrocrystalline, often just called crystalline) and very small, microscopic-sized (microcrystalline or more often called cryptocrystalline) crystals. The microcrystalline varieties are often cut and polished into ornamental stones.

There are several high-temperature and pressure structures (polymorphs) of SiO_2. Tridymite and cristobalite are found in volcanic rocks and coesite and stishovite are found as products of meteorite impact.

Macrocrystalline or crystalline varieties are differentiated by colour names:

Rock crystal: Clear, colourless quartz; milky quartz: white, translucent quartz (the most common variety); rose quartz: pale to intense pink; smoky quartz: brown to almost black; amethyst: pale to dark purple; citrine: yellow, not to be confused with the pale brown of smoky quartz.

Sometimes crystalline varieties of quartz contain another mineral, such as tourmaline or rutile, as inclusions. Tiger's eye is quartz included with fibres of the amphibole, riebeckite (p. 196) that give the specimen a chatoyant appearance.

Quartz (40677): Conchoidal fracture with one striated crystal face.
Hot Springs, Garland Co., Arkansas.
Width of specimen: 5 cm

Cryptocrystalline or microcrystalline varieties:

Chalcedony: Uniform colours of grey, blue, or brown; agate; like a chalcedony but it has various colours in it, usually in concentric bands; onyx; an agate that is banded in white and black or dark brown; jasper; opaque in shades of red, yellow, and brown; carnelian; the colours of jasper but clear; chrysoprase; translucent, apple-green chalcedony; petrified wood; a cryptocrystalline quartz in browns, reds, or yellows that has replaced wood.

Opal is an almost amorphous variety of quartz, consisting of spheres of silica gel. If these spheres are of uniform size and well-organized, they refract light into beautiful colours.

Quartz (36869): Stacked quartz crystals with hexagonal prisms and pyramids. Greely, Osgoode Tp., Carleton Co., Ontario. Width of field of view: 4 cm

Appearance

Colour: Usually white or colourless but all colours and hues are possible due to impurities.

Streak: White powder. Too hard to streak.

Lustre and transparency: Vitreous lustre and transparent to translucent.

Habit: Often as crystals in a variety of shapes: Prismatic crystals with striations perpendicular to the length and terminated with a pyramid are common. It may also be massive, granular botryoidal (rounded or globular), or stalactitic in habit.

Physical Properties

Hardness: Hard (Mohs 7).

Density: Medium (2.65 g/cm³).

Breakage: Brittle with a conchoidal fracture and no cleavage.

Test: Hardness, density, and striations on crystal faces are important. The lack of cleavage and good conchoidal fracture are noteworthy.

Similar Minerals

Calcite (p. 138) is softer (Mohs 3). It has three good cleavages and fizzes in acid. Similar differences distinguish quartz from aragonite (p. 146) and dolomite (p. 142).

Feldspar (p. 226) is slightly softer (Mohs 6) and has two distinct cleavages at nearly right angles to each other.

Nepheline (p. 232) has a greasy lustre, often with mica inclusions.

Quartz (30175): Clear, colourless, prismatic crystals with pyramidal terminations. Calcite is white rhombohedral crystals. Bluebell Mine, Riondel, Kootenay District, British Columbia. Width of specimen: 13 cm

*Quartz (38396): Flattened quartz crystal in massive quartz.
J.B. Steele Mine, Lyndhurst, Leeds & Lansdowne Tp., Leeds Co., Ontario.
Width of field of view: 8 cm*

*Quartz (40711): Doubly terminated crystal "Herkimer diamond."
Crystal Grove Campsite, Lassellsville area, Fulton Co., New York. Width of field of view: 5.5 cm*

Quartz (41424): Agate variety. Chihuahua, Mexico. Width of specimen: 4 cm

Occurrence

Expect to find quartz just about everywhere. It is most noteworthy when you don't find it, as this indicates special chemical conditions during crystallization.

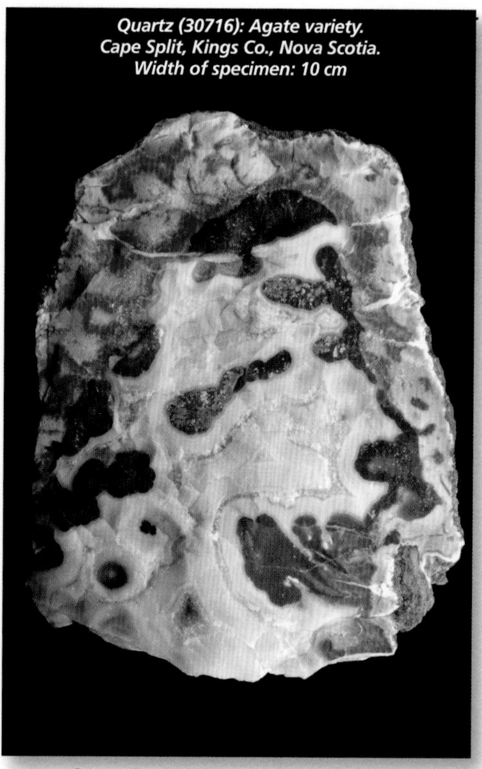

Quartz (30716): Agate variety. Cape Split, Kings Co., Nova Scotia. Width of specimen: 10 cm

Canada's Best Localities: Large, clear crystals from Bow Lake (up to 30 cm), Banff National Park, Alberta; Bluebell Mine, Riondel, Kootenay District, British Columbia; Wawa, Algoma District, Ontario; Lyndhurst, Leeds County, Ontario; Lawrenceville, Shefford County, Québec. Amethyst from Elbow Lake, Thunder Bay District, Ontario, and Bay of Fundy, Nova Scotia. Rose quartz from Quadeville, Lyndoch Township, Renfrew County, Ontario. Smoky quartz from Horn Peak, Hess Mountains, Yukon (smoky crystals); Greely, Osgoode Township, Carleton County, Ontario. Agates from Robbins Creek, Kamloops District, British Colum-

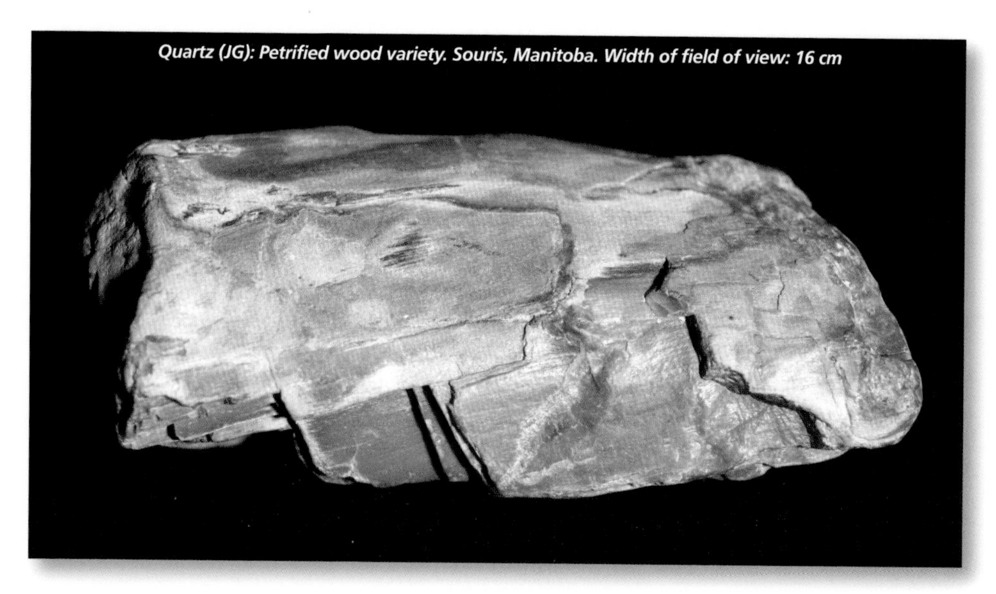

Quartz (JG): Petrified wood variety. Souris, Manitoba. Width of field of view: 16 cm

bia; Michipicoten Island, Lake Superior, Ontario; Bay of Fundy, Nova Scotia. Petrified wood from Hedly, Similkameen District, British Columbia; Red Deer River, Drumheller, Alberta; Souris, Manitoba.

Other Localities: Quartz crystals from Hot Springs, Arkansas, USA; Herkimer County, New York, USA; Minas Gerais, Brazil (up to 2 m). Amethyst from Guerrero, Mexico; Brazil and Uruguay. Smoky quartz from Brazil and Switzerland. Agate from Deschutes County, Oregon, USA; Chihuahua, Mexico; Rio Grande do Sol, Brazil; Idar-Oberstein, Germany. Petrified wood from Painted Desert, Arizona; Yakima County, Washington; Malheur County, Oregon, USA. Opal from San Luis Potosi, Mexico; South Australia and Queensland, Australia.

Quartz (51307): Amethyst variety crystals with pyramidal termination.
Diamond Willow Mine, Pearl, McTavish Tp., Thunder Bay District, Ontario.
Width of field of view: 6.5 cm

Interesting Facts

The mineral name quartz has several possible origins; *quarz* is German, *kwardy* is Slavic, and *querklufterz*, meaning cross-vein ore, is Old English.

Quartz has been an integral part of the development of physics; Pliny (23–79 CE) knew quartz split white light into a spectrum of coloured light bands; Nicolas Steno (1638–1686) studied the angles between prism faces and discovered them to always be 60º. He established the important crystallography law of constancy of interfacial angles.

Early uses included quartz wafers as a phonograph sensor, and today the same technology is applied to microbalances (scales that can weigh extremely small weights).

Charles Sawyer invented the process for producing synthetic quartz crystal and began production in Cleveland, Ohio, in 1956. This led to a huge industry in semiconductors and fibre optics.

Quartz, opal (OJ 1421): Precious opal variety. Queensland, Australia. Width of field of view: 4 cm

Quartz: Faceted citrine quartz. Sri Lanka. Width of gem: 3.3 cm

Quartz (42971): Massive rose quartz.
Quadeville Rose Quartz Mine, Quadeville, Lyndoch Tp., Renfrew Co., Ontario. Width of field of view: 15 cm

Quartz (41514): Jasper variety.
Scott Mine, Old Chelsea, Hull Tp., Gatineau Co., Québec. Width of specimen: 19 cm

FELDSPAR GROUP

Sodium, calcium potassium aluminum silicate: $(Na,Ca,K)(Si,Al)_4O_8$

The feldspar group is a fairly large group with some 20 mineral species. This group alone comprises about 60 percent of the earth's crust, hence it is important to be able to recognize these minerals. Within the feldspar group there are two sub-groups: the potassium feldspars and the plagioclase feldspars.

Common species:

Potassium feldspars with three different atomic structures:

Microcline, $KAlSi_3O_8$
Sanidine, $KAlSi_3O_8$
Orthoclase, $KAlSi_3O_8$

Plagioclase feldspars are divided along a chemical gradation:

Albite, $NaAlSi_3O_8$
Oligoclase, $(Na_{0\cdot80}Ca_{0\cdot20})AlSi_3O_8$
Andesine, $(Na_{0\cdot60}Ca_{0\cdot40})AlSi_3O_8$
Labradorite, $(Ca_{0\cdot60}Na_{0\cdot40})AlSi_3O_8$
Bytownite, $(Ca_{0\cdot80}Na_{0\cdot20})AlSi_3O_8$
Anorthite, $CaAl_2Si_2O_8$

Because of the number of similarities between species in the feldspar group, often a tentative identification is given as either plagioclase or orthoclase. Even this differentiation can be difficult. Only orthoclase or microcline can be pink and only plagioclase can be dark grey to almost black. Light-coloured feldspar with striations is plagioclase but if there are no striations, which is common, then it can be either orthoclase or plagioclase.

Feldspar (53265): Amazonite variety, consisting of two feldspars, green microcline and white albite. Kola Peninsula, Russia.Width of specimen: 8 cm

Feldspar (35804): "Perthite" variety consists of two feldspar species, orange microcline and white albite. Perth, Drummond Tp., Lanark Co., Ontario. Width of field of view: 11 cm

Appearance

Colour: Plagioclase is colourless, white, grey, dark grey, and rarely greenish or reddish. In the albite and oligoclase part of the series there are varieties called moonstone and peristerite that display an iridescent blue sheen while labradorite has the characteristic play of blue, green, and gold colours.

Feldspar (46048): Orthoclase, moonstone variety, showing a translucent blue colour and a stepped-face crystal growth. Sri Lanka. Width of specimen: 9 cm

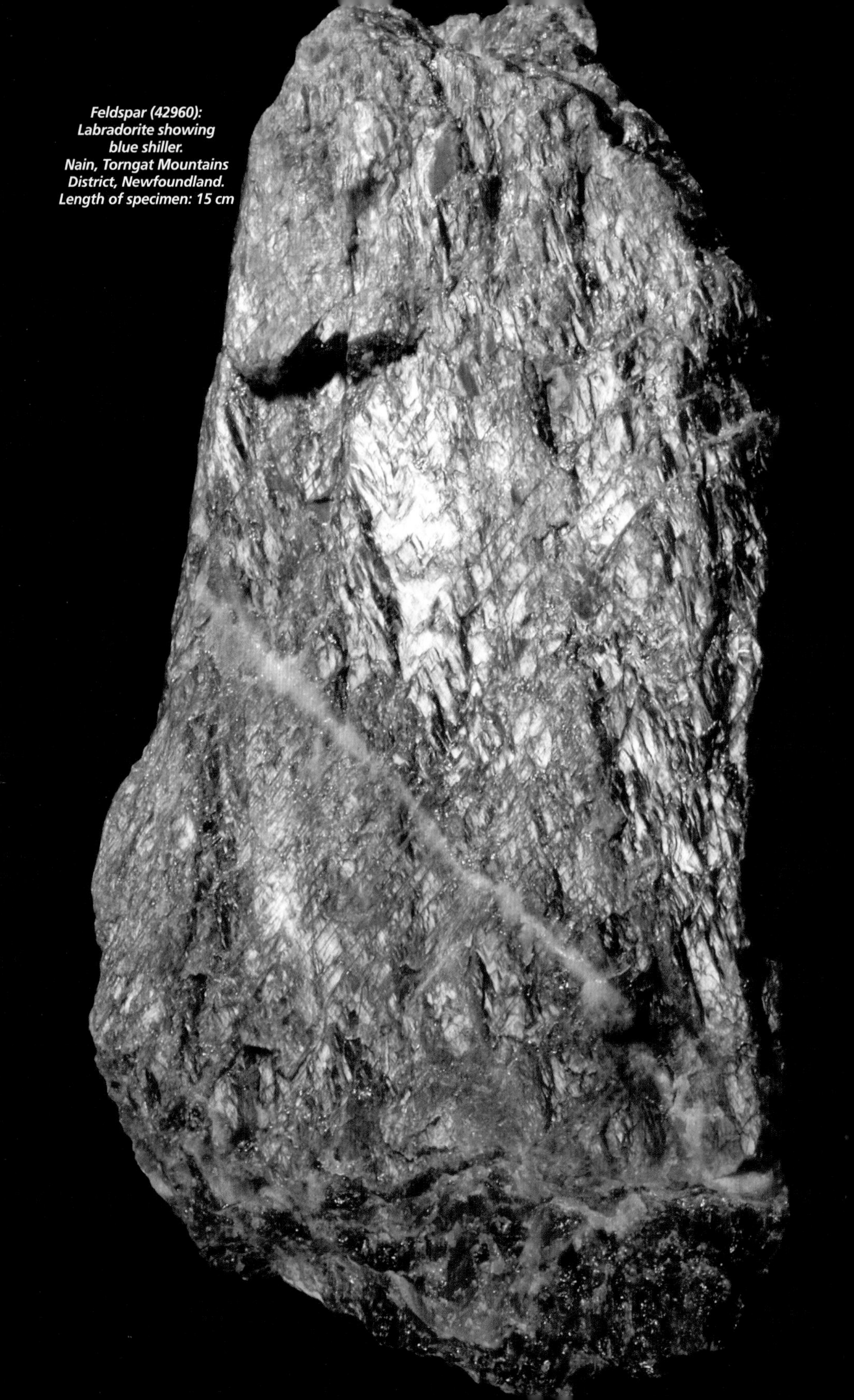

Feldspar (42960):
Labradorite showing
blue shiller.
Nain, Torngat Mountains
District, Newfoundland.
Length of specimen: 15 cm

Orthoclase is commonly a pale orange to pink colour but it is often colourless or white. Amazonite is a green variety of microcline and sanidine sometimes has the bluish tinge of moonstone.

Streak: White.

Lustre and transparency: Lustre is dull to slightly vitreous, may be pearly on cleavage faces and transparent to translucent.

Habit: Often massive or granular. Crystals occur in a variety of shapes: prismatic and tabular crystals are common. Crystals often have a banded appearance due to twin planes in the crystal growth.

Physical Properties

Hardness: Hard (Mohs 6).

Density: Medium (2.6 – 2.8 g/cm^3).

Breakage: Brittle with two distinct cleavages nearly at right angles to each other.

Test: Cleavage and hardness are important. Colour iridescence is important to differentiate some varieties.

Similar Minerals

Quartz (p. 218) is slightly harder (Mohs 7) and has a conchoidal fracture with no cleavage. Nepheline (p. 232) has only very poor cleavage and the lustre has a distinct greasy appearance.

Feldspar (51579): Microcline crystal. Poudrette Quarry, Mt. St.-Hilaire, Rouville Co., Québec. Width of specimen: 7 cm

Feldspar (45049): Oligoclase, peristerite variety, showing blue schiller. Walker Mine, Verona, Portland Tp., Frontenac Co., Ontario. Width of specimen: 6 cm

Occurrence

Feldspar is found in most rock types – both intrusive and extrusive igneous rocks and metamorphic rocks. It also forms in sediments.

In the plagioclase sub-group the identification of each species, usually by chemical analysis, is important in rock classification. In general Si-rich rocks have Na-rich plagioclase and Si-poor rocks have Ca-rich plagioclase. The amount of feldspar relative to other minerals is also important in naming rock types.

In potassium feldspars, sanidine is found in rhyolite which is a Si-rich extrusive rock that has been rapidly cooled.

Feldspar (84077): Oligoclase showing lamellar twinning planes. Soper River, Kimmirut, Baffin Island, Nunavut. Width of specimen: 7 cm

Canada's Best Localities: Sanidine and orthoclase: Beaverdell, Yale District, British Columbia; Long Lake, Monmouth Township, Haliburton County, Ontario. Microcline: Perth, North Burgess Township, Lanark County, Ontario; Maynooth, Wicklow and Monteagle Townships, Hastings County, Ontario; Mont Saint-Hilaire, Rouville County, Québec; amazonite variety from Hybla, Monteagle Township, Hastings County, Ontario; Sundridge, Strong Township, Parry Sound District, Ontario; Lac Sairs, Villedieu Township, Témiscamingue County, Québec. Albite and oligoclase: Soper River, Baffin Island, Nunavut; Balderson, Bathurst Township, Lanark County, Ontario; Purdy Mine, Mattawa Township, Nipissing District, Ontario; Davis Quarry, Dungannon Township, Hastings County, Ontario; Richmond, Melbourne Township, Richmond County, Québec. Labradorite: Nain, Eagle River District, Labrador, Newfoundland.

Other Localities: Sanidine and orthoclase: Summit County, Colorado, USA; Guanajuato, Mexico; Puy-de Dôme, France; Madrid, Spain; Val Tavetsch, Switzerland; Tyrol, Austria; Stavern, Norway; and gemmy fragments from Betroka, Madagascar. Microcline: Tveidalen, Norway and amazonite variety from Pike's Peak, El Paso County, Colorado, USA. Albite and oligoclase: St. Lawrence County, New York, USA; Narssârssuk, Greenland; Minas Gerais, Brazil; Larvik, Norway, where it is quarried for building stone (larvikite). Labradorite: Plush, Oregon, USA; Ojamo, Finland; Madagascar.

Interesting Facts

The mineral name feldspar is derived from the German, *feld*, field and *spat*, a rock that does not contain metal.

Feldspar has several important industrial uses. It is mined from pegmatites and feldspar-rich sands. Feldspar is important in making inexpensive glass. It serves as a flux to lower melting temperatures, thus saving energy. For ceramics feldspar is important in glazes and in the ceramic itself, giving it strength and durability. Feldspar is used as filler in paint, plastic, and rubber as it is stable, inert, and resistant to frost.

Feldspar varieties such as moonstone, labradorite, and amazonite are used in jewellery. They have interesting and beautiful optical properties and are relatively inexpensive.

Feldspar (46753): Labradorite, showing labradorescent colours of gold, blue, and green. Grenfell Labradorite Quarry, Tabor Island, Newfoundland. Width of specimen: 13 cm

NEPHELINE: Sodium, potassium, aluminum, silicate: $(Na,K)AlSiO_4$

Although it is not a spectacular mineral, nepheline is important in geology and in industry. It is one of a small group of minerals described as feldspathoids, meaning that it is something which is like feldspar in appearance and chemical composition but is not feldspar. The important thing in mineralogy is the ability to tell nepheline from feldspar as they represent very different kinds of rocks and hence different geology.

Appearance

Colour: Greyish white. It has been described as dovetail grey. This is accurate in that many doves have a slight bluish tint in the grey of their tails.

Streak: White.

Lustre and transparency: Greasy lustre and translucent.

Habit: Usually massive or as grains in rocks but occasionally there are crystals occurring as stubby hexagonal prisms.

Nepheline (1992.453): Rough, hexagonal prismatic crystals. Davis Hill, Bancroft, Dungannon Tp., Hastings Co., Ontario. Width of field of view: 16 cm

Physical Properties

Hardness: Hard (Mohs 5 – 6).
Density: Medium (2.6 g/cm^3).
Breakage: Brittle without a distinct cleavage but a conchoidal fracture.
Test: Colour and lustre are important. Biotite mica inclusions are also diagnostic.

Similar Minerals

Feldspar (p. 226) has two cleavages and it is more white than grey or bluish grey. The lustre of feldspar is vitreous while that of nepheline is dull and greasy.

Occurrence

Nepheline is found in igneous rocks that are low in silica. These rocks are called syenite (p. 254).

Canada's Best Localities: Crystals up to 25 cm have been found at Davis Hill, Dungannon Township, Bancroft District, Hastings County, Ontario; Nemegosenda, Chewitt Township, Sudbury District, Ontario.

Other Localities: Khibina Massif, Kola Peninsula, Russia, and the type locality at Mount Vesuvius, Italy.

Nepheline (1952.69): Massive, dovetail blue-grey colour with black mica inclusions. Davis Hill, Bancroft, Dungannon Tp., Hastings Co., Ontario. Width of specimen: 9 cm

Interesting Facts

The mineral name is from the Greek, *nephele*, meaning cloud. This description applies to the frosty appearance of nepheline when immersed in acid.

Nepheline makes better glass than feldspar as the higher aluminum content of nepheline makes a glass more resistant to scratching and breakage. The whiteness of nepheline makes it desirable in the ceramic industry, and as it melts at a lower temperature than feldspar, it helps save energy costs in both these industries.

SODALITE: Sodium, alumino-silicate, chloride: $Na_4(Si_3Al_3)O_{12}Cl$

Sodalite is close in appearance, chemical composition, and atomic structure to that of the other vivid blue mineral, lazurite. As with lazurite, the main use of sodalite is for decoration. The difference in price between rare, expensive lazurite and relatively common and inexpensive sodalite makes it worthwhile to be able to differentiate the two minerals. Sodalite can occur in non-blue varieties that are very difficult to differentiate from the closely associated nepheline.

Appearance

Colour: Usually pale to dark blue, sometimes grey-white, greenish, or reddish.
Streak: Blue streak for the blue-coloured variety.
Lustre and transparency: Vitreous lustre and transparent to translucent.
Habit: Mostly massive or as grains in a rock. Rarely it occurs as rough, rhombic dodecahedral crystals.

Physical Properties

Hardness: Hard (Mohs 6).
Density: Light (2.3 g/cm^3).
Breakage: Brittle with an uneven or conchoidal fracture.
Test: Colour and lustre are important.

Sodalite (38875): Massive with no cleavage.
Princess Quarry, Bancroft, Dungannon Tp., Hastings Co., Ontario.
Width of specimen: 10 cm

Similar Minerals

Lazurite (p. 236) has inclusions of pyrite and white calcite while sodalite will have inclusions of grey nepheline and reddish streaks. Lazurite is azure blue while sodalite is darker blue, sometimes tinted green.

Nepheline (p. 232) greatly resembles grey-white sodalite but the lustre of nepheline is greasy while that of sodalite is more vitreous.

Occurrence

It is found in syenite (p. 254) often associated with nepheline and dark mica.

Canada's Best Localities: Ice River, Kootenay District, British Columbia; Princess Quarry, Dungannon Township, Haliburton County, Ontario; crystals from Mont Saint-Hilaire, Rouville County, Québec.

Other Localities: Litchfield, Kennebec County, Maine, USA. Much of the cut and polished sodalite we see in stores today comes from Bahia, Brazil, or Walvis Bay, Namibia.

Interesting Facts

The mineral name reflects its sodium content.

Princess Quarry, near Bancroft, Ontario, is so named because Princess Mary (later Queen Mary, wife of King George V) fell in love with sodalite when she saw it on display at the Columbian Exposition in Chicago in 1893. She later requested 130 tons to be shipped from the Princess Quarry to England for use in decorating Marlborough, one of the royal residences.

Sodalite (46159): Pale-blue, massive sodalite with dark green inclusions of pyroxene (aegirine). Demis & Poudrette Quarries, Mont Saint-Hilaire, Rouville Co., Québec. Width of view: 17 cm Photo: R.A. Gault

LAZURITE (lapis lazuli): Sodium, calcium, aluminum, silicate, sulphide: $Na_3Ca(Si_3Al_3)O_{12}S$

Deep blue lazurite is a rare and expensive mineral that is popular in jewellery. Crystals are rare. It is most often found in massive form, combined with calcite and pyrite. This combination of minerals is called lapis lazuli.

Appearance

Colour: Deep blue to greenish blue.
Streak: Light blue.
Lustre and transparency: Dull lustre and translucent.
Habit: Often compact, massive. Blocky crystals with 12-sided (dodecahedral) forms are rare.

Physical Properties

Hardness: Hard (Mohs 5 – 6).
Density: Light (2.4 g/cm^3).
Breakage: Brittle with an uneven fracture.
Test: Colour is important as well as having inclusions of pyrite and white calcite.

Similar Minerals

Azurite (p. 156), another deep blue mineral, is softer (Mohs 3½ – 4) and heavier (3.8 g/cm^3). Azurite fizzes in strong acid while lazurite will not, though calcite inclusions in lazurite will fizz in acid.

Sodalite (p. 234) has a more vitreous lustre. It may have inclusions of white nepheline which is harder than the calcite inclusions of lazurite.

Lazulite, a phosphate mineral, has a similar name and colour but it has good crystals with a vitreous lustre, while lazurite is rarely crystalline and has a dull lustre.

Occurrence

Lazurite is found in sedimentary limestone rocks metamorphosed by heat (called contact metamorphism).

Canada's Best Locality: Just north of Kimmirut, Baffin Island, there is an occurrence of poor-quality lapis lazuli.
Other Localities: Lake Baikal, Russia; Badakhshan, Afghanistan.

Interesting Facts

The mineral name lazurite is either from Latin, *lazulum*, Arabic *lazaward,* or Persian *lazhuward*, meaning blue.

Lapis lazuli, also known as lapis, has a long history of use. The Egyptian pharaohs, as early as 5000 BCE, favoured it as an ornamental stone in their tombs. Their lapis originated in the northeast province of Badakshan, Afghanistan, still a source of the mineral today. Egyptian women used powdered lapis lazuli as eye shadow. In the Middle Ages and Renaissance times the tempera-paint pigment, ultramarine, was derived from grinding lapis lazuli and removing the impurities until it was pure lazurite. This highly expensive pigment was replaced by cobalt blue and shortly thereafter synthetic ultramarine, both in the 19th century.

As with many precious minerals of adornment there is inevitably an enhancement or synthetic invented. The most common enhancement is to take poor-quality lapis lazuli, impregnate it for durability, and stain it a darker blue. The newest imitation, the Gilson product, comes complete with pyrite specks but no calcite. It is of interest that cobalt blue, ultramarine, and Gilson imitation were all invented in France.

Lazurite (31268): Granular.
Soper River, Kimmirut, Baffin Island, Nunavut.
Width of specimen: 11 cm

ZEOLITE GROUP

Potassium, sodium, calcium, aluminum silicate, water: $(K,Na,Ca)_2[(Si,Al)_8O_{16}]\cdot 6H_2O$

This is a large mineral group with almost 50 species. They are not just beautiful minerals to collect, they also have important applications in agriculture, industry, and health. The group is quite rare in nature but huge amounts of them are made synthetically to satisfy industry needs. Their usefulness is a result of an open, cage-like crystal structure.

The more common zeolite species are:

Analcime: $NaAlSi_2O_6\cdot H_2O$; white; trapezohedral (12 kite-shaped faces) crystals.

Chabazite: $CaAl_2Si_4O_{12}\cdot 6H_2O$; white, yellowish, orange; rhombohedral crystals (6 diamond-shaped faces).

Heulandite and clinoptilolite: $(Na,Ca)_2Al_2Si_4O_{12}\cdot 6H_2O$; white, yellowish, orange; pearly lustre; tabular with diamond-shaped faces).

Mordenite: : $(Ca,Na)\ Al_2Si_{10}O_{24}\cdot 7H_2O$; colourless to white, fibrous.

Natrolite: $Na_2Al_2Si_3O_{10}\cdot 2H_2O$; colourless or white; long prismatic to needle shaped.

Scolecite: $CaAl_2Si_3O_{10}\cdot 3H_2O$; colourless or white; long prismatic to needle shaped.

Stilbite: $NaCa_4Al_9Si_{27}O_{72}\cdot 30H_2O$; white, yellowish; complex bundles of crystals.

Appearance

Colour: Usually colourless or white but also pale shades of orange, green, yellow, pink, or brown.

Streak: White.

Lustre and transparency: Vitreous to pearly lustre, and transparent to translucent.

Zeolite (37917): Pale orange, rhombic chabazite and trapezohedral, white analcime in vug. Nova Scotia. Width of specimen: 12 cm

Zeolite (36994): Natrolite prism.
Demix & Poudrette Quarries, Mt. St.-Hilaire, Rouville Co., Québec.
Width of field of view: 5 cm

Zeolite (83827): Stilbite ball on chabazite.
Maniwaki area, Angoumois Tp., Pontiac Co., Québec.
Width of field of view: 7.5 cm

Zeolite (40574): Yellow sheath of stilbite
with small clear calcite crystals.
Nova Scotia.
Width of field of view: 6 cm

Habit: Often as crystals in a variety of shapes: prismatic, needle-shaped and tabular crystals are common.

Physical Properties

Hardness: Moderate (Mohs 3 – 5).
Density: Light (2.2 g/cm^3).
Breakage: Brittle with a distinct cleavage in some species and an uneven fracture.
Test: Mode of occurrence and density are important. Heating causes the mineral to boil off its water.

Similar Minerals

Within this large group it is difficult to tell many species apart but the important factor is to recognize it as a zeolite group mineral.
Quartz (p. 218) is harder (Mohs 7) and heavier (2.65 g/cm^3).
Feldspar (p. 226) is harder (Mohs 6) and heavier (2.7 g/cm^3).

Occurrence

The zeolite group minerals are found in volcanic basalt, usually in cavities. Large deposits of clinoptilolite are formed as a result of volcanic ash being altered at low temperatures in the presence of water.

Canada's Best Localities: Crystals of stilbite, natrolite, chabazite, and heulandite can be found in the basalts of the Bay of Fundy, Nova Scotia, and New Brunswick. Natrolite

Zeolite (84010): Stilbite "fan" with pearly lustre. Stilbite. Maniwaki, Egan Tp., Gatineau Co., Québec. Width of field of view: 5 cm

Zeolite (39092): Mordenite needles in basalt. Monte Lake area, Kamloops Division, Yale District, British Columbia. Width of field of view: 8.5 cm

crystals from the nepheline syenites of Mount Saint-Hilaire, Québec, and Zinc Mountain, Ice River area, British Columbia.

Other Localities: Natrolite crystals from Lane County, Oregon, and Somerset County, New Jersey, USA. Fabulous crystal groups come from several localities; Nasik, Bombay, India; Strontian, Scotland; Faeroe Islands; and Iceland.

Interesting Facts

The name zeolite is from the Greek, *zein*, to boil, and *lithos*, meaning stone or rock. This well-chosen name refers to the important property of zeolites of readily releasing the weakly-bonded water in their structure when mildly heated.

There are two crystal structure features that make zeolites very useful. The open, cage- or channel-like framework can be used as a micro-filter for purifying or extracting substances that are very small, such as removing bacteria in water purification, removing smelly sulphur dioxide from nat-

Zeolite (32126): Spray of natrolite needle-shaped crystals. Scots Bay, Kings Co., Nova Scotia. Width of field of view: 6.5 cm

ural gas, and producing oxygen by separating it from water. These channels can also act as a catalyst, promoting chemical reactions such as producing gasoline from oil.

The second important structural feature of zeolite minerals is their loosely-bonded, large atoms like sodium, calcium and potassium. These may be replaced (cation exchanged) to absorb and retain other compounds, including radioactive waste, oil spills, and cat urea.

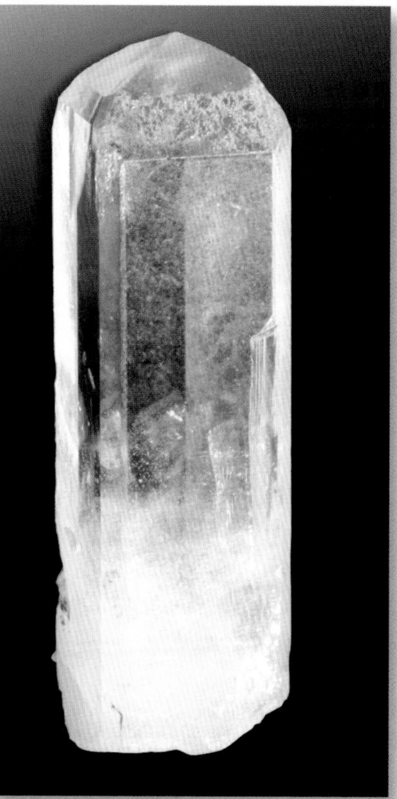

Natrolite crystal (44004): With orthorhombic prism and pyramidal termination. Ice River Complex, Zinc Mt. Ridge, Moose Creek Valley, Kootenay District, British Columbia. Length of crystal: 4 cm

Zeolite (30710): Heulandite crystals with pearly lustre and lozenge shape. Big Eddy Slide, Cape Blomidon, Kings Co., Nova Scotia. Width of field of view: 8.5 cm

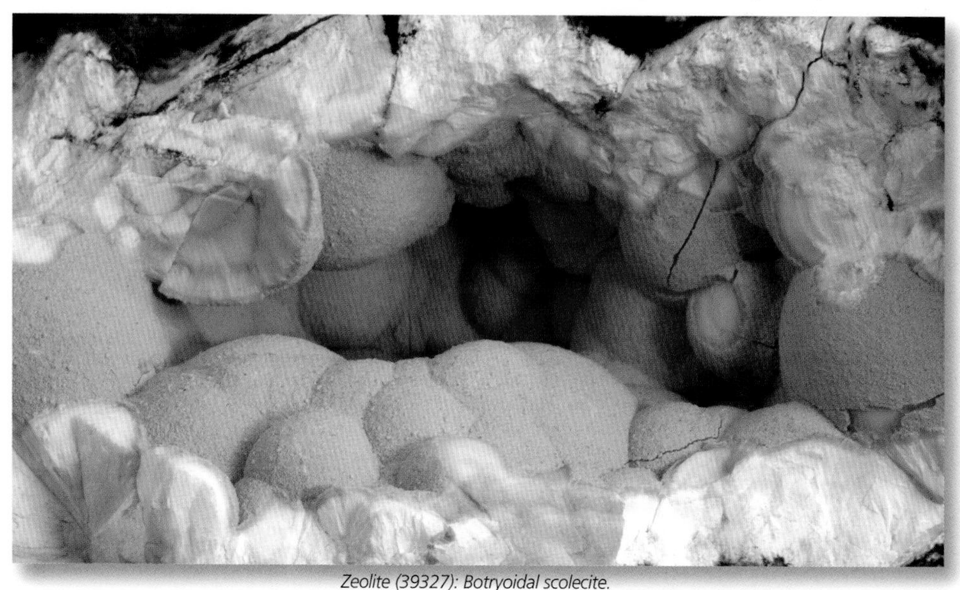

Zeolite (39327): Botryoidal scolecite.
Jeffrey Mine, Asbestos, Shipton Tp., Richmond Co., Québec. Width of field of view: 6 cm

ROCKS

GRANITE

Born of molten magma deep inside the earth, granite is the most common igneous or plutonic rock. It's composed of three main ingredients: quartz, feldspar, and mica. The several varieties of granite are named based on colour (red granite, white granite); texture of crystals (graphic granite, orbicular granite); or on the accessory or minor mineral content (hornblende granite). Use the proportions found in the Mineral Content section to identify the rock. The figures show the typical proportions one sees in granite.

Appearance

Colour: Depends on the quantity and variety of feldspars, quartz, mica, and other minor minerals. White to light grey, yellowish, pink and red are possible colours of feldspar. Quartz is a glassy, smoky grey to white colour. Black specks can come from biotite mica or hornblende, while brownish to silvery specks can come from muscovite mica.

Texture: Typically randomly arranged mineral grains and can be coarse to very coarsely grained. Particularly coarse-grained granites are called pegmatites.

How Formed

Granite forms deep inside the earth's crust as cooling magma crystallizes. Slow cooling of the magma produces large crystals, while fast cooling produces small ones. Granite forms during periods of mountain building.

Canadian occurrence: Québec is noted for its extensive granite quarries.

Granite, graphic intergrowth (R3): Microcline feldspar is green with dark-coloured quartz. Bouchette, Quebec. Width of field of view: 9 cm

Granite, pink (R599): Orange microcline feldspar with clearer quartz. St. George, Charlotte Co., New Brunswick. Width of field of view: 7 cm

Mineral Content

Granite contains 60 percent feldspar, at least 20 percent quartz, and 10 percent micas. This includes the varieties of feldspars, such as microcline, orthoclase, and albite; and the micas biotite and muscovite. Minor amounts of hornblende, augite, and magnetite may also be present.

Similar Rocks

Granite gneiss (p. 290) is a metamorphic rock showing prominent layers or bands that result from its formation under extreme pressure.

Diorite (p. 250) is darker coloured as it contains a larger proportion of dark green to black minerals.

Interesting Facts

As a major component of the earth's crust, granite is highly resistant to erosion, making it a prominent feature in many landscapes. This hard, weather-resistant rock makes steep, sheer cliffs that are sought out by mountaineers and rock-climbers.

Uses: Building materials and monuments. "Black granite," as a label on stone carvings in art galleries is usually gabbro (p. 252) or diorite (p. 250).

Granite pegmatites are a source of the gems tourmaline and topaz.

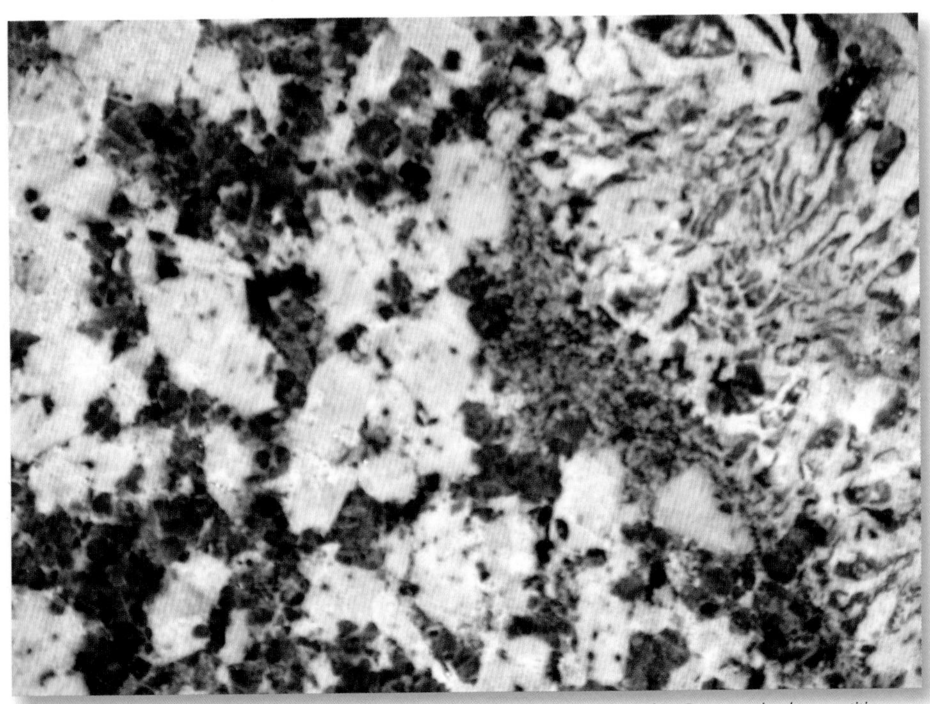

Granite grey (R252): Light-coloured albite feldspar with dark smoky quartz grains. Coarse-grained pegmatitic. Wolf Lake occurrence, Seagull Creek, Yukon. Width of field of view: 6 cm Photo R.A. Gault

Granite pegmatite (81619): Pale, pink microcline feldspar; white, platy albite feldspar (top right); smoky quartz; green, prismatic, elbaite tourmaline. Virgen da Lapa, Minas Gerais, Brazil. Width of specimen: 19 cm Photo: R.A. Gault

DIORITE

Diorite has a chemical composition and appearance intermediate between granite and gabbro. It may be described as a "salt and pepper" rock due to the random mixture of dark- and light-coloured minerals. Diorite may be associated with either granite or gabbro intrusions and may grade into either. It is the chemical equivalent of the extrusive, igneous rock andesite (p. 260). Diorite is associated with ore deposits that are rich in amphibole minerals such as the huge porphyry copper deposits of the southern USA.

Appearance

Colour: Diorite is medium to dark grey. Even the calcium-rich plagioclase feldspar in diorite is dark.

Texture: Medium to coarse-grained rock with crystals or grains all the same size.

How Formed

Diorite results from solidifying magma deep in the earth's crust (i.e., an intrusive rock). The magma is poorer in silica (SiO_2) and richer in iron and aluminum oxides compared to granite. It is commonly formed in volcanic arc areas and in mountain building regions. Diorite can occur in huge volumes, in batholiths.

Mineral Content

Diorite is essentially plagioclase feldspar and hornblende amphibole. Sometimes there may be some pyroxene, biotite mica, or quartz.

Interesting Facts

Diorite is a very hard rock, thus it takes a high polish and is durable. Because of these qualities, diorite has been used since ancient times in Egypt. The code of Hammurabi was inscribed in approximately 1790 BCE in Mesopotamia (Middle East) on a two-metre pillar of diorite. (Some references give the rock type as basalt.)

Uses: In many European cities durable and resistant diorite was, and to a limited extent still is, used as cobblestones. Although rougher in texture than asphalt, cobblestones have several environmental advantages: they are much more weather- and abrasion-resistant, thus lasting longer and not releasing undesirable hydrocarbons when taken into solution. They can also be pulled up for work on electrical conduits or water pipes, then replaced.

Due to the hardness of diorite it is seldom carved but it is excellent for polishing and engraving. It is used in monuments, cemetery headstones, flooring, and walls. It is often sold under the name "black granite" which of course is quite misleading and inaccurate.

Similar Rocks

Gabbro (p. 252) is also a dark-coloured igneous rock but it contains no quartz, more pyroxene, and much less amphibole than diorite.

Granite (p. 248) is lighter in colour than diorite due to its large content of feldspar and quartz.

Diorite (R1000): Light-coloured plagioclase with dark green amphibole. Highland Bell Mine, Beaverdell, Similkameen Division, British Columbia. Width of field of view: 7.5 cm

GABBRO

Gabbro is a dark, basic (very little silica in the composition) rock in which quartz is very rare. It forms large igneous bodies at depth within the earth and is only exposed at the surface after millions of years of erosion.

Appearance

Colour: Gabbro is dark green to almost black with creamy white patches. When it has been weathered it is brownish.

Texture: Coarse-grained with individual grains being a few millimetres in size.

How Formed

Gabbro forms from a magma that is rich in iron and magnesium, and poor in silica (quartz). The magma cools and crystallizes deep below the earth's surface. Gabbro has the same chemical composition as basalt and diabase, but because it cools more slowly, it grows larger crystals.

Mineral Content

Gabbro contains plagioclase feldspar, augite pyroxene, olivine, and sometimes magnetite.

Interesting Facts

The name "gabbro" is derived from a town in the Tuscany region of Italy where it was first described. The oceanic crust, which is a large proportion of the earth's crust, is gabbro.

Uses: Gabbro is another rock commonly referred to as "black granite." It is used as a building stone as well as for countertops and cemetery headstones.

Gabbro is a source of many important elements such as chrome and titanium. The huge deposits of copper, nickel, and platinum at Lynn Lake and Thompson, Manitoba, and Sudbury, Ontario, are associated with gabbroic rocks.

Similar Rocks

Basalt (p. 262) is about the same colour as gabbro but much finer grained.
Diorite (p. 250) is lighter in colour and may contain more quartz and lighter coloured feldspars.
Serpentinite (p. 298) is much finer grained than gabbro and has no distinct crystal outlines.

Gabbro (R571): Light-coloured plagioclase with dark green-brown pyroxene.
Label indicates location is either Larvik, Norway, or Lac St. Jean, Québec.
Width of field of view: 7 cm

SYENITE

Syenite is light-coloured and hence is often confused with granite. It is important to differentiate these two rock types as each represents a very different mode of formation. Syenite originates from magma containing a lot of aluminum oxide (Al_2O_3) while granite formation depends on a large amount of silica (SiO_2).

Appearance

Colour: Light-coloured, usually white, grey, tan, or yellow.

Texture: Coarse-grained rock in which individual grains or crystals can be differentiated with the unaided eye.

How Formed

Syenite, like granite, forms deep in the earth's crust (termed a plutonic igneous rock). The basic chemical composition of the two rocks is fundamentally different, with syenite having more aluminum oxide (Al_2O_3) and granite having more silica (SiO_2).

Mineral Content

Syenite contains mostly feldspar which is rich in sodium (Na) and potassium (K). Sometimes minor amounts of quartz are present but never more than 10 percent of the total mineral content. Syenite may also contain a considerable amount of nepheline, in which case it is called nepheline syenite.

Interesting Facts

Nepheline syenite often contains exotic minerals that are of interest to mineral collectors and scientists.

Uses: Syenite is commonly quarried for road aggregate and as a raw material in the production of cement.

Similar Rocks

Granite (p. 246) has a minimum of 20 percent quartz while syenite rarely has quartz at all and it never exceeds 10 percent.

Syenite (R1004): Keremeos, Similkameen Division, British Columbia.
Width of field of view: 6 cm

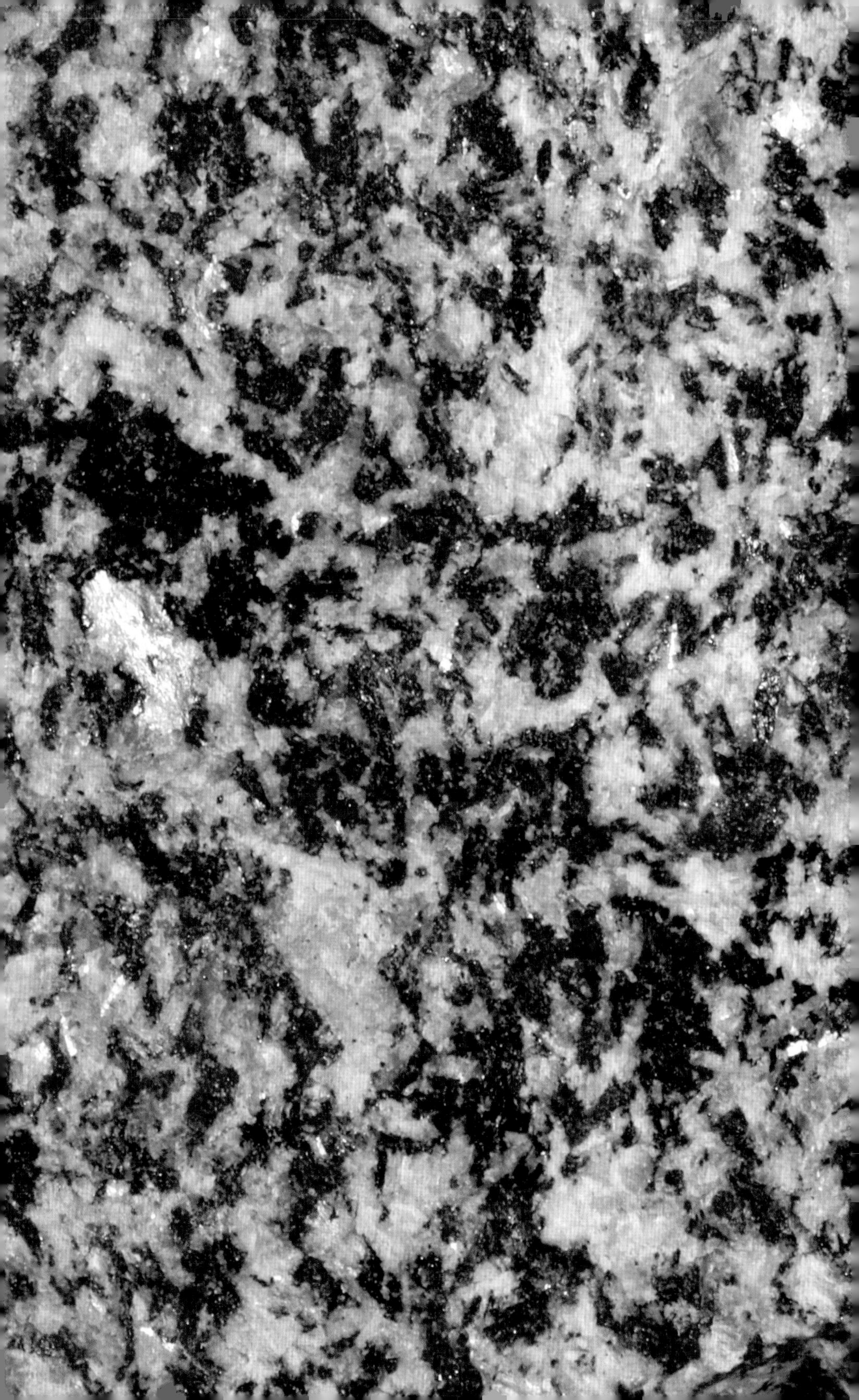

RHYOLITE

Rhyolite is the surface equivalent of granite, which forms deep in the earth. It is a widespread volcanic rock that is common at recent eruptions around the globe. The eruptive event is often explosive and accompanied by large volumes of volcanic ash (pumice) dispersed to enormous heights and covering wide areas.

Appearance

Colour: Usually light-coloured – grey, tan, or yellow, sometimes with a reddish, greenish, or brownish tint.

Texture: Fine-grained rock usually with no visible crystals. Occasionally fine crystals are visible; in this case the rhyolite is called porphyritic. Rhyolite may also have cavities or vesicles left by gas bubbles.

How Formed

Rhyolite forms in volcanoes. When the lava is extruded, it freezes or solidifies quickly. The majority of the resulting rhyolite is glass, much like that of basalt but the composition is more like granite: silica-rich (SiO_2) with less sodium (Na), calcium (Ca), magnesium (Mg), and iron (Fe). Pumice is a type of rhyolite that is full of vesicles.

Mineral Content

Rhyolite is largely glass but it may contain microscopic crystals of quartz, feldspar, and often biotite mica.

Test: Because of its glass content, rhyolite breaks with a ringing sound and it is very rough to touch. Pumice, being full of vesicles, can float in water.

Interesting Facts

Rhyolite lava is very viscous or thick because of the high silica content. This viscous lava may plug the volcanic vent hole, causing pressure to build and resulting finally in a very explosive eruption.

Rhyolite scoria (R659): Scoria with gas bubbles and ropy texture.
Near Klamath Falls, Klamath, Oregon.
Width of field of view: 22 cm

Rhyolite comes from the Greek, *rhyo*, meaning to flow. The apparent contradiction to its high viscosity as a lava is probably due to the phenomenon known as *nueé ardent*. This "burning cloud" moves rapidly when the eruption is supercharged with gas that lowers the viscosity of the rhyolite.

Examples of rhyolite flows are: The disastrous eruption of Mount Pelée that buried the town of Saint Pierre, Martinique, in 1902 and killed over 30,000 people. The flows of Rotorua volcanic area, New Zealand, cover some 25,000 square kilometres. Vast areas of western North America are also covered by these huge rhyolite flows.

Uses: Rhyolite itself is of little use, except for minor amounts that are powdered for use as a scouring agent. However, this glassy material is the country rock of several important ore deposits such as gold from Rhyolite, Nevada, copper ores of Noranda, Québec, and the Potasi silver area in Bolivia.

Similar Rocks

A very fine-grained sedimentary sandstone (p. 272) or metamorphic quartzite (p. 296) resembles rhyolite, but rhyolite has more glass and less crystals.

Andesite (p. 260) is darker in colour.

Rhyolite (R477): Known locally as "Picture Stone." These patterns are caused by diffusion of water and colourants through the rock. Hope Slide, near Hope, British Columbia. Width of field of view: 16 cm Photo: R.A. Gault

Rhyolite pumice (R658): Eruptive ash forming rhyolite pumice with fine-grained glassy fragments. Millard Co., Utah. Width of field of view: 5 cm Photo: R.A. Gault

ANDESITE

Andesite is a rock composed mainly of plagioclase feldspar (andesine or oligoclase). It is a volcanic extrusive rock comparable in chemical composition to the intrusive rock, diorite. Andesite is between basalt and rhyolite in chemical composition for extrusive rocks. Be careful not to confuse andesite, the rock name, with andesine, the feldspar mineral name.

Appearance

Colour: Usually medium grey but can shade to almost black.

Texture: Fine-grained rock sometimes having clots of light-coloured feldspar crystals (termed porphyritic).

How Formed

Andesite results from lava flows or extrusions from deep-seated magmas having a diorite composition. It is the most common volcanic rock at the margins of continental plates. An example of this would be the "ring of fire" that surrounds the Pacific Ocean and is delineated by a series of active volcanoes of andesite composition.

Andesite (941): Andesite with dark, needle-like amphibole (hornblende) included crystals. Mount Shasta, California.Width of field of view: 7 cm Photo: R.A. Gault

Mineral Content

Plagioclase feldspar (andesine or oligoclase) is the most significant constituent but pyroxene, amphibole, and biotite mica may be present in lesser amounts.

Interesting Facts

The word andesite is derived from the Andes Mountains located along the west coast of South America, where this rock type is very common.

The very viscous nature of silica (silicon-dioxide-rich lavas) such as andesite leads to highly explosive and spectacular eruptions. In 1883 the island of Krakatau, near Jakarta, Indonesia, erupted in a series of cataclysmic explosions that could be heard at a distance of over 4,600 kilometres and sent ash to a height of 11 km. Two thirds (roughly 23 square km) disappeared into a caldera and slipped below sea level resulting in a tsunami about 40 m high. Some 36,000 people lost their lives.

The famous 1902 eruption of Mount Pelée, on the Caribbean Island of Martinique, was no less spectacular. It began with a series of earth tremors and a showering of ash. These events caused 30 cm centipedes and poisonous snakes to leave the slopes of Mount Pelée and enter the nearby town of St. Pierre. The centipede bites were painful and the snake bites fatal to some 50 persons. But this was nothing compared to the events of the following days. The finale was marked by a 150 kilometre-per-hour flow of superheated gas, ash, and rock. It buried St. Pierre in minutes, killing 28,000 people and leaving only two survivors.

Uses: As road aggregate it is used to build the road base. It is also quarried for building stone. Geologists are interested in andesite not as the source of economic mineral deposits but as the host. Gold, silver, and base metals (copper, lead, and zinc) are mostly found in altered andesite.

Similar Rocks

Rhyolite (p. 256) is fine-grained but lighter in colour.

BASALT

Basalt is a dark-coloured rock that is the extrusive equivalent of gabbro. Since it is volcanic and flows out onto the surface of the earth, it cools quickly and thus is composed largely of glass because minerals do not have sufficient time to crystallize out of the liquid. It is the most abundant volcanic rock.

There are several types of basalt that are named for their dominant physical feature. Vesicular basalt has holes (vesicles) in it as a result of gas bubble formation before "freezing." If these vesicles are filled with minerals such as zeolites or agate quartz, it is termed amygdaloidal basalt. Some basalt has begun to crystallize and the crystals appear as clots of light-coloured feldspar or dark green olivine. These clots are termed phenocrysts and the rock is known as porphyritic basalt.

Appearance

Colour: Dark grey to black, often with a very dark green tint.

Texture: Very fine-grained, largely glass with crystals that are invisible or barely visible with a hand lens.

Basalt (73-1-1): Typical fine-grained, dark basalt.
Near Actinolite, Elzevir Township, Ontario. Width of specimen: 16 cm
Photo: R.A. Gault

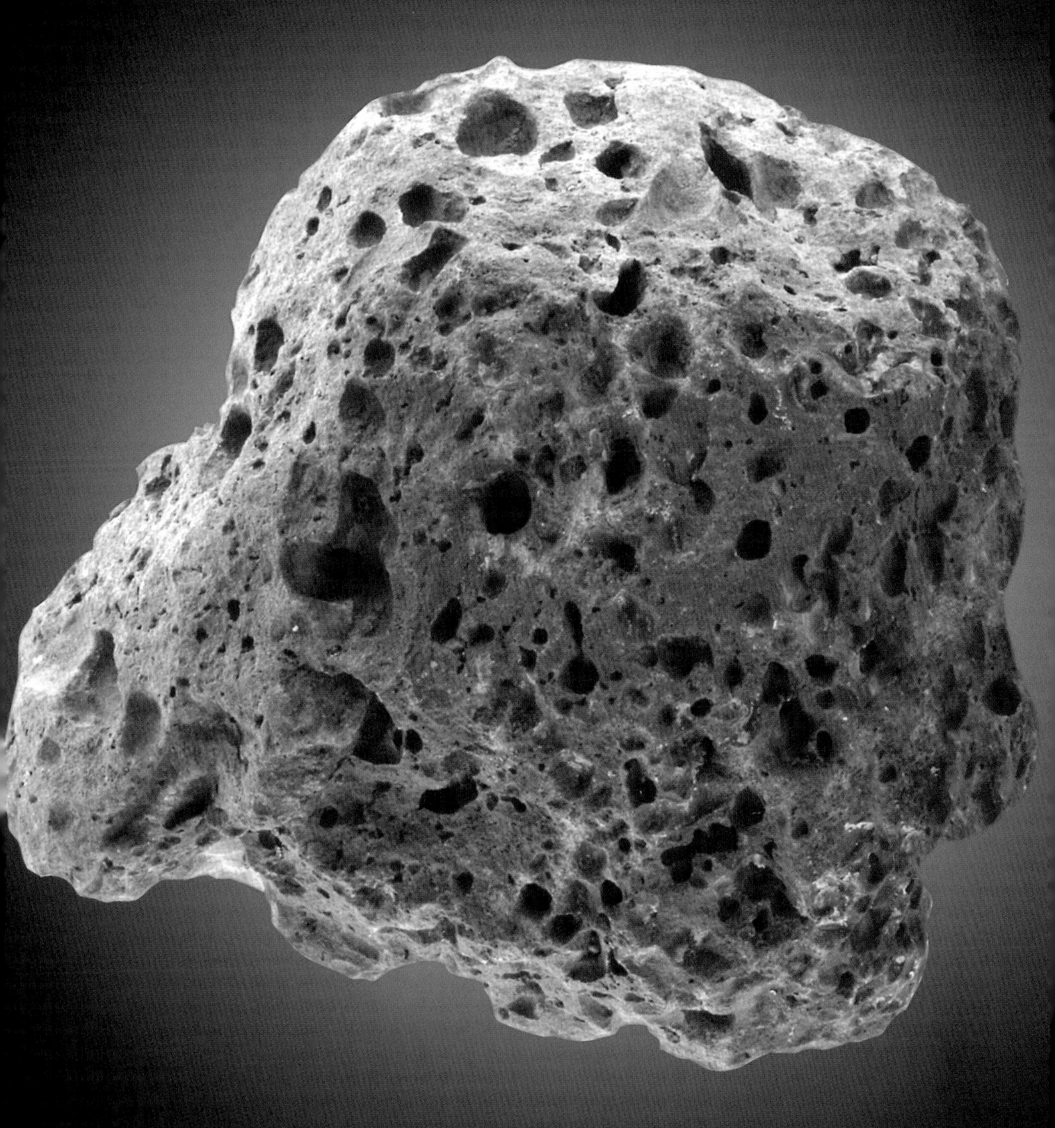

Basalt, vesicular (R698): Vesicular basalt.
East of Princeton, British Columbia.
Width of specimen: 15 cm

How Formed

Basalt is a silica-poor volcanic extrusive rock. It comprises most of the ocean floor.

Mineral Content

It is largely glass with small crystals of calcium-rich plagioclase feldspar, olivine, and pyroxene.

Test: When you break basalt with a hammer it sounds like breaking glass.

Interesting Facts

Flood basalts extrude from huge cracks in the floor of the ocean. Individual flows may be up to 300 m thick and cover thousands of square kilometres. Northern Ireland's Giant's Causeway is an example of how basalt can fracture into pentagonal-shaped columns upon cooling.

Basalt is a source of agate, zeolites, and gemmy olivine.

Uses: Crushed basalt is used in building construction and road engineering. A recent development is "basalt fibre." Basalt rock is quarried then melted and extruded into fine fibres. Basalt fibres are considered superior to other fibres in durability, thermal stability, electrical, sound and heat insulation, and vibration and chemical resistance. These properties make it useful in plastic pipes, brake pads, fire-retardant materials, and reinforcement in composites used in automotive parts and boat hulls.

Similar Rocks

Sedimentary rocks such as dark siltstone (p. 274) may resemble basalt, but when hit with a hammer it does not have the ring that basalt does.

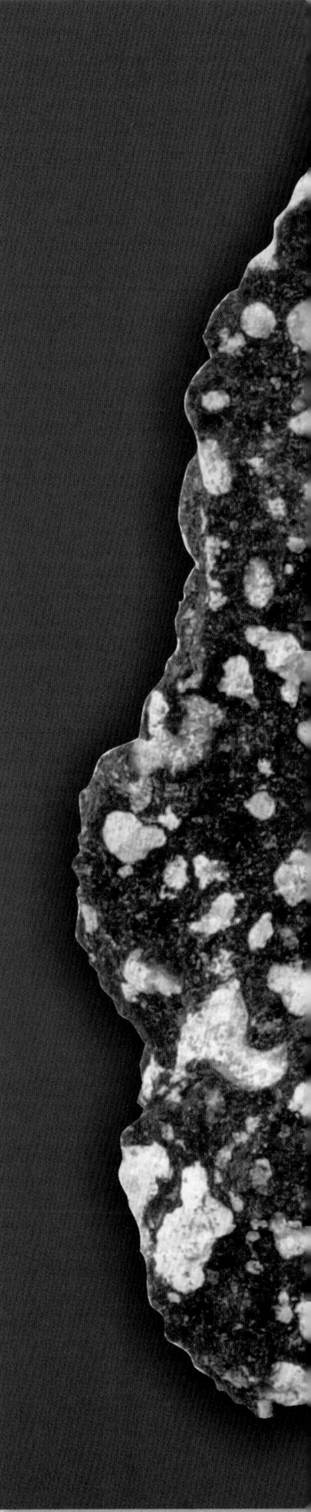

Basalt, amygdaloidal (R694): Basalt with amygdules filled with zeolites. Batch Bay, Lake Superior, Ontario. Width of specimen: 11 cm

OBSIDIAN

Obsidian is a unique rock as it is actually glass – a substance with no crystal structure. This naturally occurring glass forms at the edges or surface of rhyolite volcanic lava flows where it cools so rapidly that there is no time for crystallization. The temperature of formation would be about 1000ºC.

Appearance

Colour: Usually black with occasional brownish, greenish, or greyish tints. It may include swirls of red colour. Sometimes it has white patches of fine crystals resembling snowflakes (snowflake obsidian). These snowflakes are the mineral cristobalite, a high temperature polymorph of quartz.

Texture: A glass that is massive but occasionally teardrop-shaped (with the variety name Apache tears). A broken surface will have a conchoidal (scoop-shaped) fracture, often with ribs or ridges. Obsidian breaks into shards with very sharp edges.

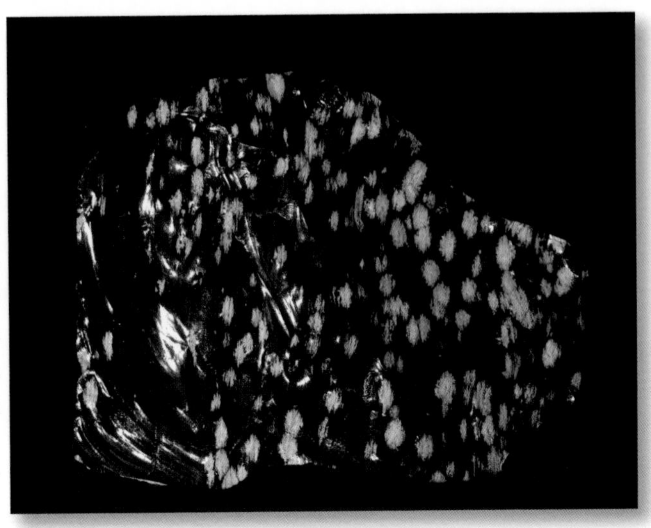

Obsidian, snowflake (R612): Snowflake obsidian displaying conchoidal fracture. The "snowflakes" are minute feldspar crystals. Nephi, Utah. Width of specimen: 14 cm Photo: R.A. Gault

How Formed

This glass results from a volcanic eruption that has cooled so quickly there has been no opportunity for crystals to grow. It is said to be "quenched." Obsidian has a lot of silicon oxide (silica) in its chemical composition. The dark colour is from impurities of iron. The "snowflakes" are cristobalite.

Mineral Content

Obsidian is a glass, so strictly speaking it contains no minerals, since minerals by definition are crystalline. An exception is the snowflake variety.

Test: The glass breaks like bottle glass into shards and sharp edges. The conchoidal fracture is an important diagnostic.

Interesting Facts

Obsidian has a chemical composition close to that of granite or rhyolite. With about 70 weight percent silica (SiO_2) the dark colour is from lesser amounts of iron and magnesium. It has a

Mohs hardness of 5–5½ and a density of 2.6 g/cm³. Obsidian has less than 2 weight percent water. If the glass contains more than 2 weight percent water it forms a slightly different rock called perlite.

Uses: Obsidian's tendency to form very sharp edges upon breaking has led to its use for cutting tools since Palaeolithic times. There is evidence of widespread practical and ritual use of obsidian throughout Pre-Columbian (14,000 BCE to 1500 CE) Mesoamerica.

Even today the sharp, cutting edge of obsidian is utilized in precision scalpels for heart and eye surgery. Their sharpness is superior by many times that of surgical steel, as obsidian can be honed to near atomic levels of thinness.

A minor amount of obsidian is also cut and polished for jewellery.

Obsidian fragment (R565): Obsidian glass is translucent with conchoidal fracture and knife-edge sharpness. Near Prineville, Crook Co., Oregon.Width of field of view: 15 cm

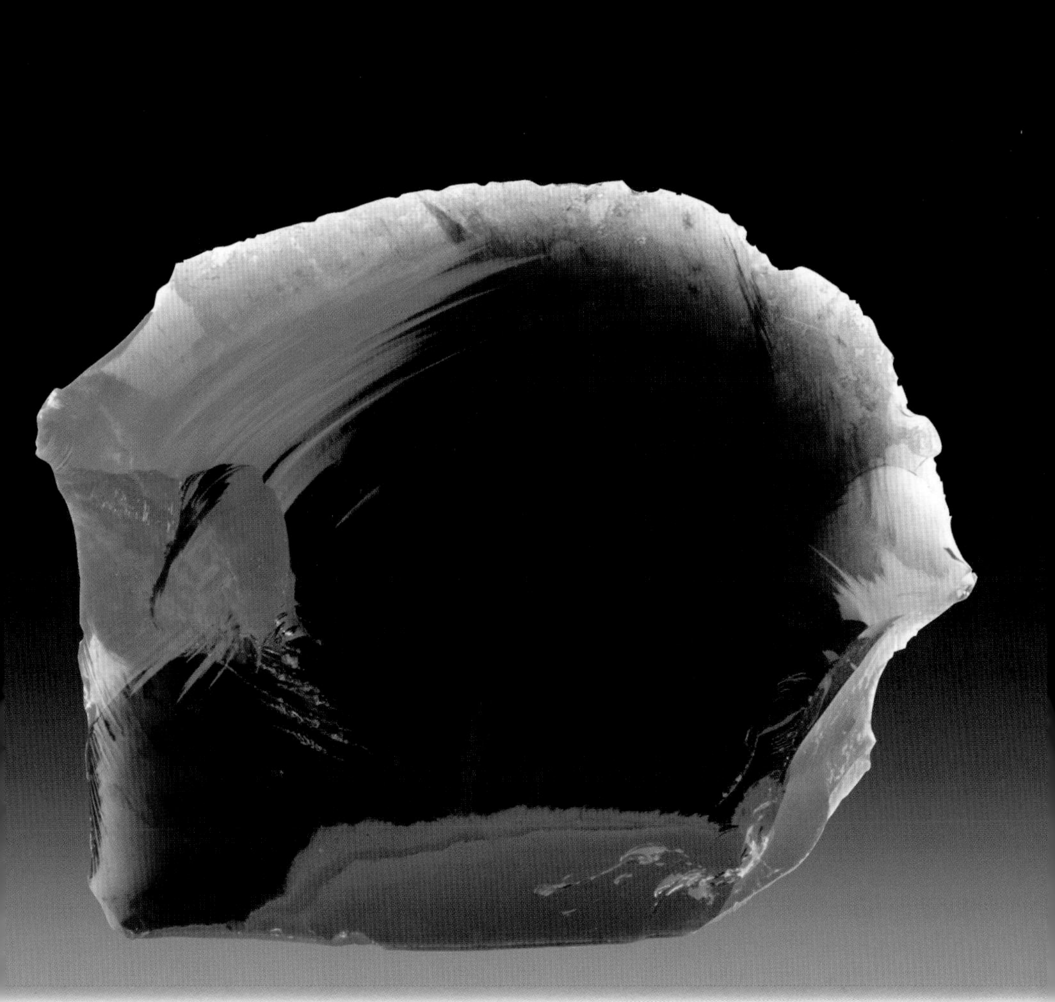

CONGLOMERATE

Conglomerate is an easy rock to identify on the whole, but its pebble constituents are not so easy as there could be a wide range of rock types. A conglomerate represents a turbulent deposit that has not had the time, or the environment was too restricted, to sort the composite material carefully. It is what is termed "poorly sorted." What you see in a conglomerate represents a sample of whatever was in the region being actively eroded at the time.

Appearance

Colour: Conglomerate usually has a light-coloured matrix with any number of inclusions of sand grains and rounded pebbles.

Texture: Fine, to coarse-grained inclusions in a fine matrix.

How Formed

Conglomerate forms in an energetic environment like a river that flows with sufficient strength to tumble, erode, and deposit large, rounded pebbles. Between the pebbles, sand forms a matrix and cement is provided from mineralized water solutions seeping through the mixture.

Mineral Content

Conglomerate contains mostly quartz as this mineral is resistant to weathering. In some cases the pebbles could be a great mixture of mineral and rock types, but those most resistant to abrasion will be the only ones to survive such rigorous transport.

Test: The most diagnostic feature is the rounded pebbles that may be of a wide range of size and composition.

Interesting Facts

There are some famous ore deposits in conglomerate rocks. These are of two types. In those like the Rand in South Africa, the gold would have been deposited as a placer in the conglomerate and possibly redistributed slightly afterward. The Blind River, Ontario, uranium deposits probably have similar origin. In the case of the huge native copper deposits of Michigan, USA, it is evident that the conglomerate merely provided a porous passage within which hydrothermal solutions passed (permeated) and deposited copper from the solution.

"Basal conglomerate" denotes a sedimentary rock lying on a lower rock that has no relationship to the conglomerate. It is an important geological feature that marks a break in time and the beginning of a new sedimentary period of deposition.

Similar Rocks

Breccia (p. 270) has angular fragments, not rounded pebbles like those in conglomerate.

Conglomerate (R295): Conglomerate with rounded pebbles cemented together.
Brackley Beach, Queen's Co., Prince Edward Island.
Width of field of view: 24 cm
Photo: R.A. Gault

BRECCIA

Breccia is a rock composed of angular fragments of any rock type or mixture of rock types. These jagged and blocky fragments of igneous, sedimentary, or metamorphic rock may be of a mixed variety of sizes, all cemented together.

Appearance

Colour: Usually a light-coloured matrix but the fragments could be of any colour or composition.

Texture: Fine- to medium-grained matrix with angular fragments of any size.

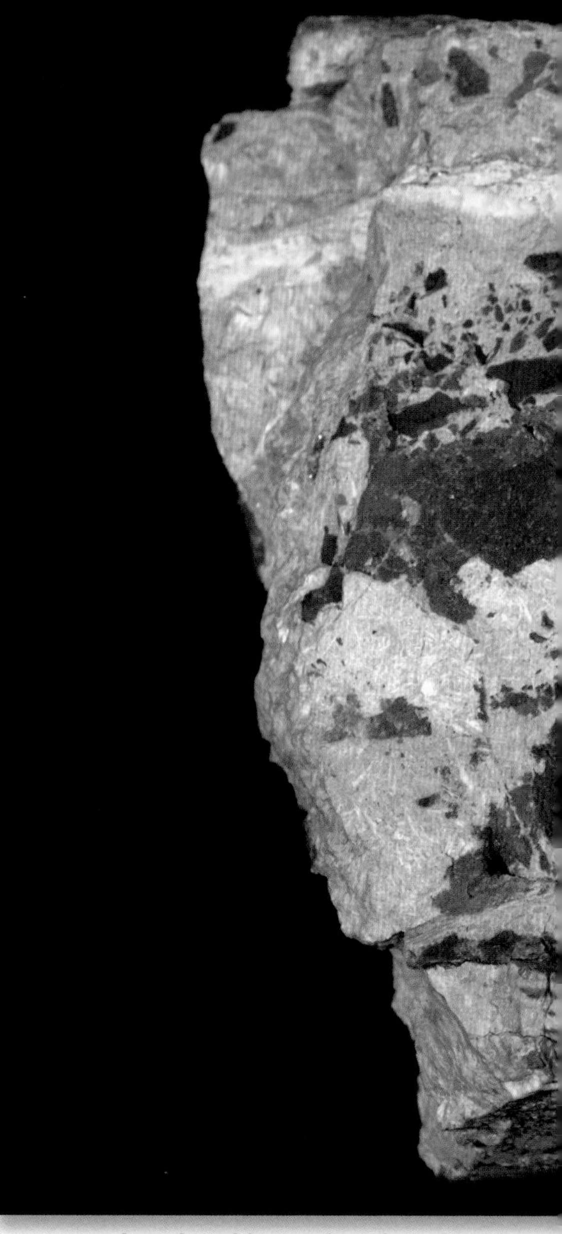

How Formed

Breccia results from rapid erosion with little transport of the fragments, thus they remain jagged. An example of such an active environment would be at the base of a cliff, a scree slope. Broken rock piles up, the weight eventually compresses the fragments together, and circulating, mineralized waters provide the cement.

Mineral Content

The fragments in a breccia can comprise a wide range of rock types or mineral species but within a particular breccia the fragments tend to be of one type since transport is minimal.

Interesting Facts

Continental edges are marked by breccia formation as two tectonic plates grind and crumble at the junction of their movements. This can also occur on a smaller size in fault zones. Another interesting source of breccia is from a meteorite impact that shatters the rock and the fragments are consolidated later in geological history.

Uses: Breccias are quite attractive, hence their use as ornamental stone since the time of the ancient Egyptians.

Similar Rocks

The inclusions in a conglomerate (p. 268) are rounded pebbles whereas in breccia they are jagged, broken fragments.

Breccia (R251): Breccia with jagged fragments cemented together. Carleton Co., Ottawa, Ontario. Width of field of view: 27 cm Photo: R.A. Gault

SANDSTONE

Sandstone is a general term for rock that is mainly composed of quartz. The varieties of sandstone depend on colour and mineral content. Try the acid or vinegar test (see pg. 4) to determine the presence or absence of calcite in the matrix cement. Sandstone may form in a large number of environments (freshwater lakes or rivers, sea beds or shores, wind or glacial deposits). The mineral content, matrix, shape of the grains (rounded or angular), and sorting of the grain size all contribute to the story of how this rock came to exist.

Appearance

Colour: Sandstone is often red or brown. Sometimes it is light grey to almost white if it is very pure. It can also have a greenish or yellowish tint. The yellow, red, and brown colours are due to the presence of iron oxides. Green colour is probably due to coloured mica. Sandstones may be colour-banded.

Texture: Usually sandstone is a medium-grained rock with fragments that are rounded and all about the same size (termed "well sorted").

How Formed

Weathering of rocks produces quartz sand that may be deposited by water or wind as sand bars, beaches, deltas at the mouths of rivers, or dunes. The deposits are compacted by further burial and cemented by mineral-rich waters that seep through it. The cement may be calcite, clay, or silica.

Mineral Content

Sandstone contains mostly quartz (p. 218) and sometimes minor amounts of feldspar and mica. If the feldspar reaches amounts exceeding 25 percent it is called "arkose."

Test: Look for individual, rounded, or angular grains. You may need a magnifying lens.

Interesting Facts

Sandstone formations are important aquifers that allow the percolation of water through the somewhat porous structure of the rock. This type of aquifer is better at filtering pollutants than an aquifer dependent on permeation through fractured rocks such as limestone.

The beauty of sandstone has been captured in nature at Arches National Park, Utah, USA, and in construction of parts of the Taj Mahal, India.

Uses: Sandstone is a prime building material. It is also used in grindstones and sharpening stones.

Similar Rocks

Siltstone (p. 274) is fine grained.

Quartzite (p. 296) is more compact and harder to break with a hammer.

Sandstone (R638): Fine-grained sandstone.
Prince Edward Island.
Width of field of view: 21 cm

SILTSTONE or MUDSTONE

Siltstone is a general term for rocks that are mainly composed of quartz. The varieties depend on origin and mineral content. In some areas, calcareous materials are deposited together with quartz and feldspar. Since this rock usually is cemented by calcite, try the vinegar test to determine the presence or absence of calcite. Traditionally mudstone contains more clay than siltstone but the two are difficult to tell apart.

Appearance

Colour: Usually light-coloured, grey, tan, or yellow, but sometimes contains impurities that make it red or black.

Texture: Fine-grained rock in which individual grains cannot be seen. It may be layered or laminated due to variations in silt size, organic matter, or calcite or dolomite content.

How Formed

Siltstone results from a compaction of fine silt fragments in both marine and freshwater environments. It is said to be "immature" since it has not weathered long enough for the feldspar in it to decompose to clay, leaving just clay and quartz.

Mineral Content

Siltstone contains mostly quartz (p. 218) but feldspar may be common and sometimes there are minor amounts of calcite or dolomite, clay, and organic material.

Test: It has a gritty feel even when wet. It may fizz slightly in weak acid due to calcite or dolomite content.

Interesting Facts

Microscopic fossils may be used to determine mode of formation, whether in marine or fresh water.

Uses: Sharpening stones.

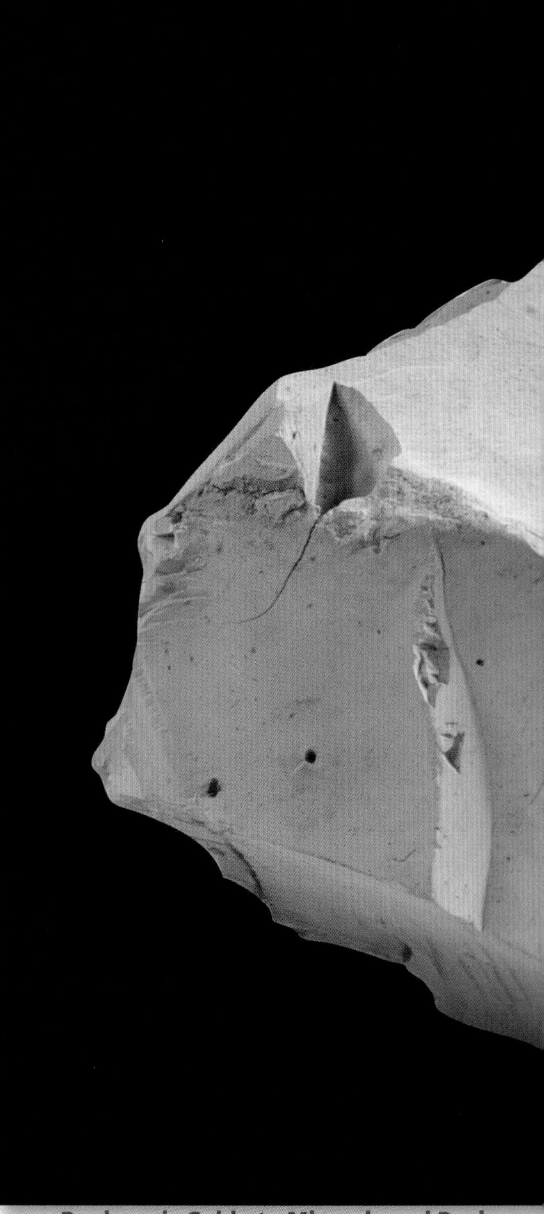

Similar Rocks

Shale (p. 282) is softer and slippery when wet due to its clay content.

Slate (p. 284) has a parting or cleavage due to aligned mica during compaction.

Siltstone (R682): Very fine-grained siltstone with a few larger grains included. Ottawa, Carleton Co., Ontario. Width of specimen: 18 cm

LIMESTONE

Limestone is a general term for rock that is composed mainly of calcite. The varieties of limestone depend on its origin, minerals, and fossil content. Most commonly, it forms in shallow seas from deposits of the remains of plants or animals, like giant cemeteries of accumulated sea life. In some areas, the calcareous material is deposited together with sand and mud, forming a composite rock like argillaceous limestone or calcareous sandstone. Since this rock is mainly composed of calcite, try the vinegar test to determine the presence or absence of calcite.

Appearance

Colour: Usually white, grey, tan, or yellow, but sometimes contains impurities that make it red or black.

Texture: Fine to medium-grained rock that often contains marine fossils.

Limestone, fossil (R803): Fossil brachiopod limestone. North America.
Width of field of view: 7.5 cm Photo R.A. Gault

Limestone: Tan coloured, fine-grained.
Width of field of view: 6 cm
Photo: R.A. Gault

How Formed

Limestone results from a chemical reaction in sea water forming a lime mud that sinks to the bottom and consolidates. Limestone can also be deposited from solutions in warm or cool waters in caves. It forms stalactites like icicles from the ceiling and stalagmites like pillars on the floor. Travertine limestone, resembling Swiss cheese, is a decorative rock formed as a cave fills with deposits. Rarely, limestones form in freshwater but only fossil remains could indicate such an origin.

Mineral Content

Limestone contains mostly calcite (p. 138) and sometimes minor amounts of dolomite, quartz, and feldspar.

Test: The powdered rock fizzes in white vinegar.

Interesting Facts

Limestone is an essential part of the carbon cycle (CO_2). It is both a source and a reservoir for CO_2. The amount of CO_2 in the atmosphere is determined partly by the amount of limestone exposed to weathering on the Earth's surface. When sea levels are low, more limestone is exposed, atmospheric CO_2 increases, and the climate warms. This is one example of how rocks and minerals affect life on Earth.

Uses: Extensively used in construction as a building stone and as an essential ingredient of cement.

Tyndall stone is a limestone that comes from a quarry north of Winnipeg, Manitoba. Examples are found in the Manitoba Legislative Building and the Canadian Parliament buildings in Ottawa.

Similar Rocks

Marble (p. 294) has larger, visible crystals.

Dolostone (p. 280) looks essentially the same but contains more dolomite. Crushed dolostone does not fizz when vinegar is applied.

Limestone, pisolitic (R423): Ibex Hills, San Bernardino County, California. Width of field of view: 8 cm Photo: R.A. Gault

DOLOMITE or DOLOSTONE

Dolomite is the name of a mineral and a rock. As a rock, dolomite consists of grains of dolomite mineral and other minerals such as quartz and calcite cemented together. Some geologists refer to the rock as "dolostone" to avoid confusion with the mineral name.

Appearance

Colour: Dolomite is usually grey, tan, or creamy brown.
Texture: It is a fine- to medium-grained rock.

How Formed

Dolomite forms in marine environments. Some geologists speculate that dolomite is a chemical replacement of limestone.

Dolostone (R220): Curved, ping crystals with younger, colourless, complex crystals on top. St.-Eustache Quarry, St.-Eustache, Deux-Montagnes Co., Québec. Width of field of view: 5 cm

Mineral Content

Dolomite contains mostly the mineral dolomite (p. 142) and sometimes minor amounts of calcite, quartz, and feldspar.

Test: Effervesces slightly in acid.

Interesting Facts

Research is ongoing to determine the exact origin of dolomite. It is a very common rock in geological history and represents about 10 percent of all sedimentary rocks, yet there are no modern formations of this rock. In the laboratory it requires minimal temperatures of 100°C to synthesize dolomite but geological evidence indicates that many vast formations must have formed at temperatures much lower than this. One suggestion based on present day observations is that dolomite will form at low temperatures in the presence of sulphate-reducing bacteria.

An important occurrence of dolomite is the Niagara Escarpment in southern Ontario.

Uses: Dolomite from a quarry on the Bruce Peninsula (part of the Niagara Escarpment) was used to construct the Canadian Embassy in Washington, D.C. Other famous buildings made from dolomite are the Royal Ontario Museum, Toronto, and the Whitney Block and Macdonald Block in the Legislative Buildings, Toronto. Dolomite may be used instead of limestone in the manufacturing of cement or as a flux in smelting iron ore. It is also used to reduce acidity in soils and neutralize acids used in paper production.

Dolomite is often the reservoir for oil.

Similar Rocks

Limestone (p. 276) is usually lighter in colour than dolomite and often contains fossils, which dolomite does not.

Sandstone (p. 272) will not effervesce in acid to the same extent as dolomite.

SHALE

Shale is a general term for rocks that are mainly composed of clay particles. The varieties of shale depend on colour and mineral or fossil content. Shale is quite a common rock and its presence can tell us a lot about conditions that existed in times past.

Appearance

Colour: Shale may be red, black, brown, dark green, or even bluish. Organic matter tends to make shale black whereas iron, in various oxidation states, makes it red or green.

Texture: Shale is very fine-grained and individual particles cannot be seen with the eye or a magnifying lens.

How Formed

Shale formation indicates a quiet, stable water environment. It may form in marine or freshwater situations such as lagoons, deep offshore marine environments, or quiet lakes. The presence of pyrite indicates shale formation in reducing (no oxygen) conditions.

Mineral Content

Shale contains mostly clay (p. 212) and sometimes minor amounts of calcite, dolomite, quartz, and feldspar. Pyrite or fossils may also be present.

Test: Wet shale feels slippery due to the clay content and it smells like mud. A fingernail can scratch or gouge wet shale.

Interesting Facts

The world famous Burgess shale near Field, British Columbia, contains fossils that are 540 million years old and represent life in an equatorial sea at that time.

Uses: During shale formation in certain areas there was a considerable amount of organic debris, primarily algae. Petroleum from these rocks may be generated by distillation at approximately 500ºC. However, the process is costly and creates environmental problems.

Shale (pyrite 46449): Brachipod shell replaced by pyrite in black shale. Mt. St.-Bruno area, Chambly Co., Québec. Width of field of view: 5 cm

Similar Rocks

Siltstone (p. 274) is harder. When siltstone is wet it feels more gritty than shale.

Slate (p. 284) has lamellar cleavage due to the formation of mica upon compaction.

Shale, oily.
Manuels River, Harbour Main Co., Newfoundland.
Width of field of view: 5 cm
Photo: R.A. Gault

SLATE

Slate is the first step in regional metamorphism, forming at lower pressures and temperatures than schist. The main indicator that this rock has been metamorphosed is the fissile splitting along planes. In some slate, aligned crystals of mica or chlorite can be seen.

Appearance

Colour: Usually black, grey, sometimes with a greenish or bluish tinge. It can also be red or reddish brown.

Texture: Fine-grained with evident smooth, flat layers. Slate readily splits along these planes (termed "fissile") and individual plates can be quite thin (termed "slatey").

How Formed

Slate results from a low-grade regional metamorphic event. Shale or clay sediments are subjected to mild heat and pressure on a large scale. Fossils may be preserved but distorted.

Mineral Content

Slate contains mostly clay minerals (p. 212) and varying amounts of quartz, feldspar, and mica or chlorite (p. 208). Mica and chlorite align perpendicular to the direction of the regional pressure. This alignment of planar minerals gives slate its perfect cleavage. Pyrite (p. 46) can form perfect crystals during slate formation.

Test: The rock readily splits along planes into thin plates.

Interesting Facts

Golden pyrite crystals contrasted with black slate make beautiful collector's items.

Uses: Slate is used for flooring tiles and roofing shingles.

Similar Rocks

Shale (p. 282) and siltstone (p. 274) do not have the characteristic splitting into sheets that slate has. Slate is harder than either shale or siltstone.

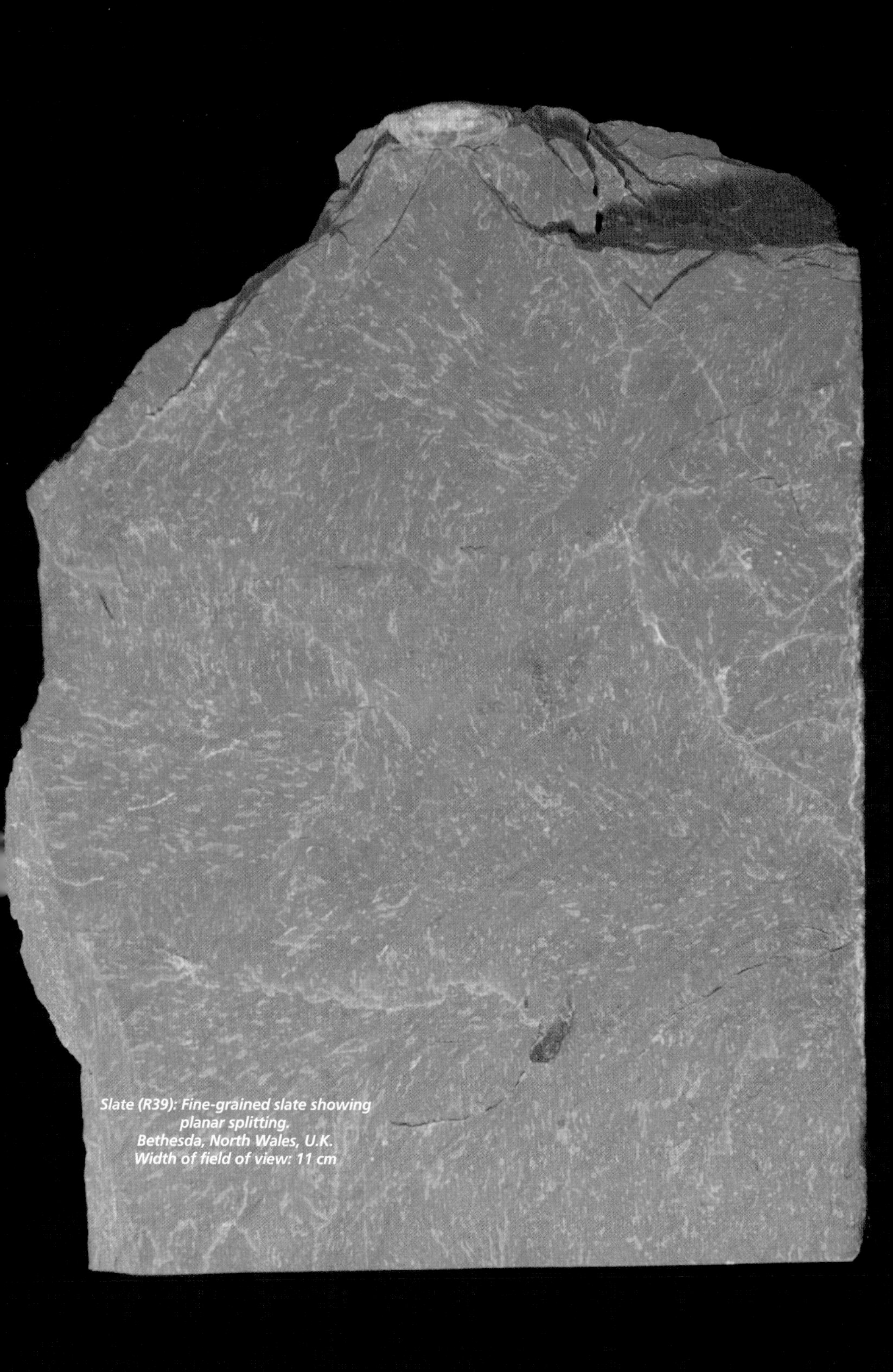

Slate (R39): Fine-grained slate showing planar splitting.
Bethesda, North Wales, U.K.
Width of field of view: 11 cm

SCHIST

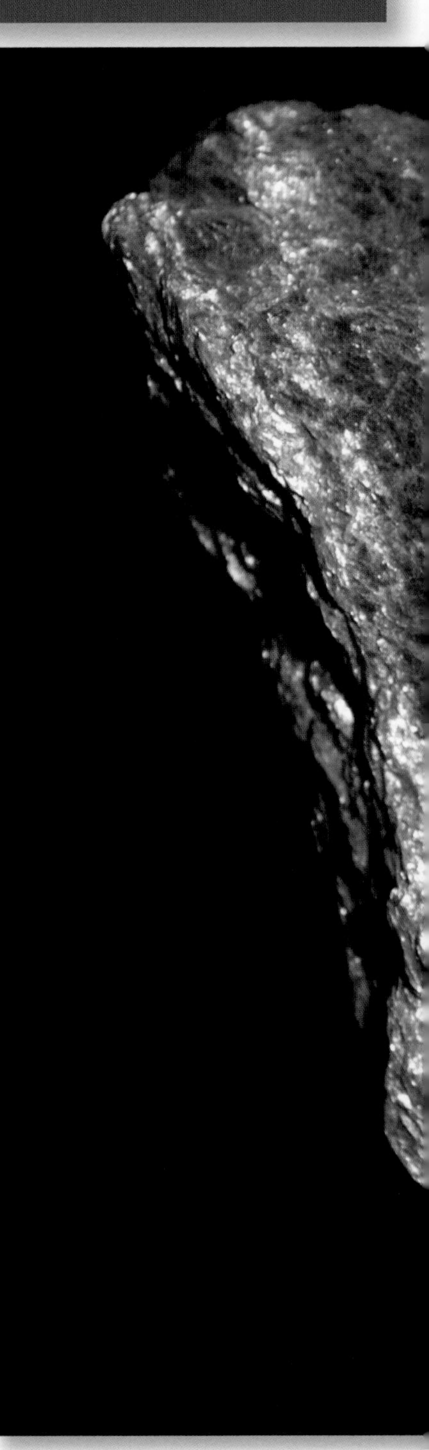

Schist (pronounced sh-ist) is a common, regionally metamorphosed rock indicative of moderate temperatures and pressures such as a mountain-forming event. A schist rock has been subjected to a higher degree of metamorphism than slate but to lesser degree than gneiss. Schists are named according to their most significant mineral content; i.e., muscovite schist, biotite schist, garnet schist, or kyanite schist.

Appearance

Colour: Schist can be a variety of colours ranging from silvery white muscovite schist to darker, almost black biotite schist.

Texture: Schist is a medium-grained rock that breaks readily with a wavy or uneven surface (termed "schistocity"). The schistocity is due to the alignment of micas into planes as a result of the pressure it has been subjected to. These planes of mica give schist its typical and diagnostic "schiller" or bright reflection.

How Formed

Schist results from the effects of a moderate pressure and low to moderate temperature on a large, regional scale. The initial rock is usually sedimentary but volcanic rocks could be involved as well. The bulk chemical composition does not change but the temperature and pressure effects are sufficient to grow new minerals such as mica, garnet, or kyanite.

Mineral Content

Schist typically contains mica, quartz, and feldspar. There are many other possibilities for inclusion such as hornblende amphibole or actinolite amphibole, kyanite, and almandine garnet. By definition there are more than 50 percent platy or elongate minerals in schist.

Test: Schistocity or breaking along wavy planes of medium to coarse mica is diagnostic.

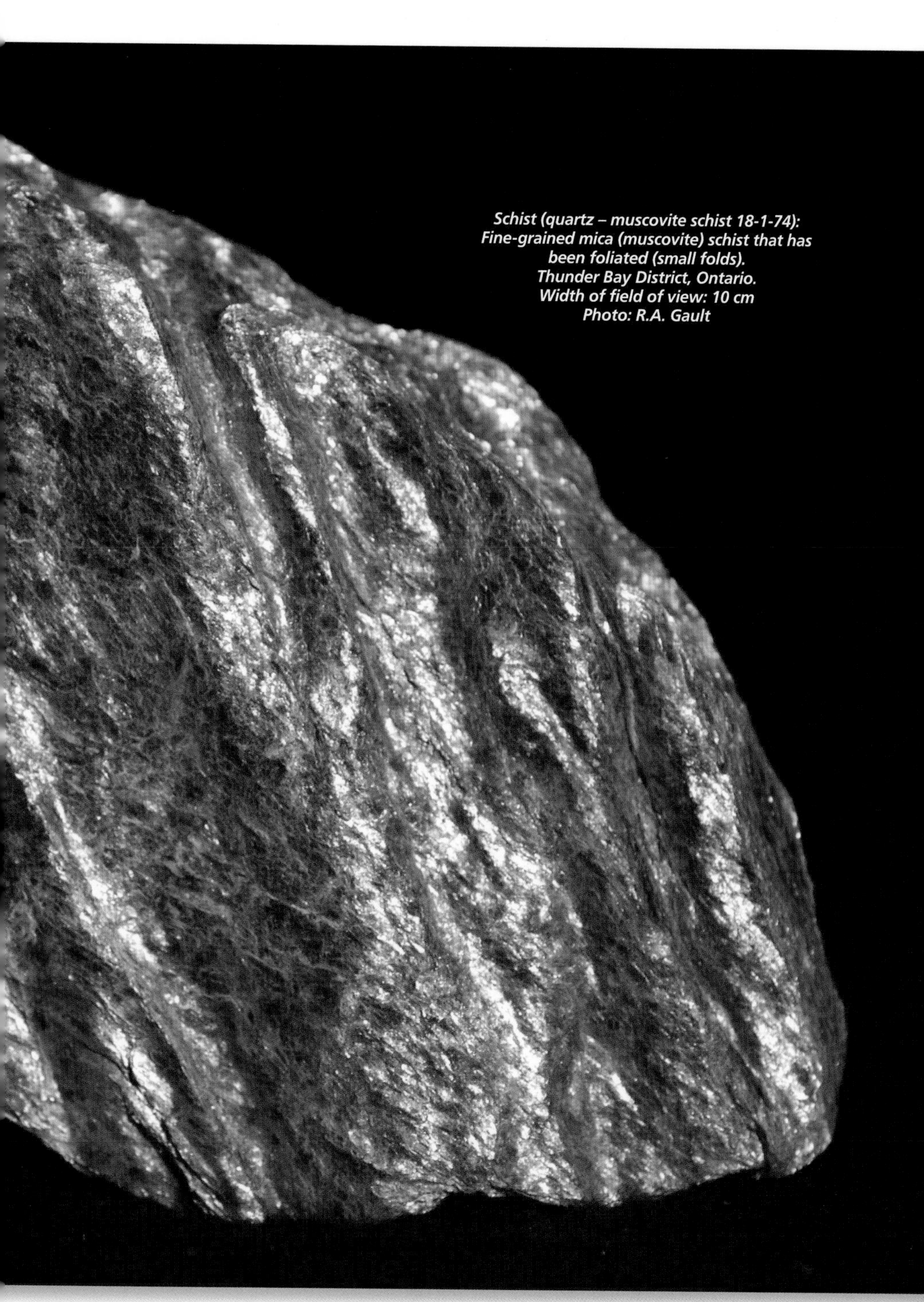

Schist (quartz – muscovite schist 18-1-74):
Fine-grained mica (muscovite) schist that has
been foliated (small folds).
Thunder Bay District, Ontario.
Width of field of view: 10 cm
Photo: R.A. Gault

Interesting Facts

The word schist is derived from the Greek, *schistos*, to split.

Uses: A commercial source of garnet for abrasives. Schist rocks have been used in many buildings but their inherent platy nature makes them susceptible to water saturation and subsequent weakness.

Similar Rocks

Slate (p. 254) is finer grained and tends to split into thin, flat plates whereas schist is coarser grained and splits into wavy plates.

Gneiss (p. 290) does not split along planes like schist does. The reason for this is that in schist the mineral alignment, particularly mica, is planar, whereas in gneiss the banding of minerals is not continuous.

Schist (R994): A quartz-muscovite schist. The muscovite sheets "sparkle." Thunder Bay District, Ontario. Width of field of view: 12 cm

Schist, garnet, mica (R146): Chisel Lake, Snow Lake area, Manitoba.
Width of field of view: 10 cm Photo: R.A. Gault

GNEISS

Gneiss (pronounced nice) is a general term for rock that has undergone high temperature and pressure metamorphic changes. The word gneiss comes from an old Slavonic term meaning *nest*. The "nest" refers to the layers or foliations in the rocks that enclosed ores being mined in medieval times. Gneiss is applied to metamorphic rocks that have undergone more extreme events than a schist rock. The term does not imply a specific mineralogy or composition but describes a texture that is an expression of a high degree of metamorphism.

Appearance

Colour: Usually light in colour but darker gneisses do exist.

Texture: A medium- to coarse-grained rock with distinct layering or banding, marked by the segregation and alignment of light and dark minerals. The light and dark layers or bands are not continuous but tend to be interrupted into a series of dashes. In some instances knots or eyes

Gneiss, granitic, augen (R343): This is termed an augen gneiss because of the lenticular-shaped orange feldspar. Grass Lake, south of South Hammond, New York, USA. Width of specimen: 20 cm
Photo: R.A. Gault

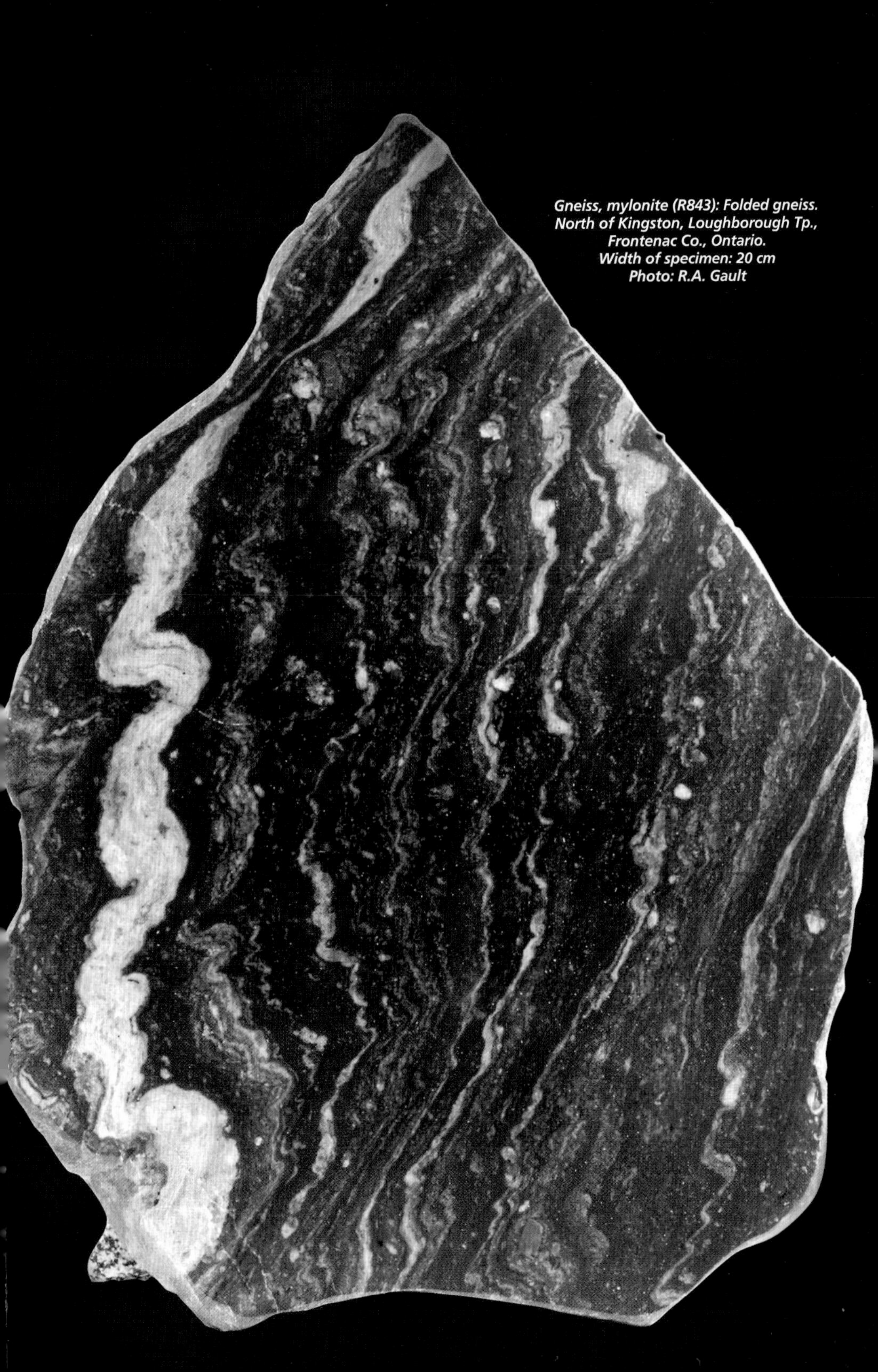

Gneiss, mylonite (R843): Folded gneiss.
North of Kingston, Loughborough Tp.,
Frontenac Co., Ontario.
Width of specimen: 20 cm
Photo: R.A. Gault

are segregated in the rock. This is termed augen, meaning eye, gneiss because the knots are lenticular, or lens-shaped. Gneiss is often folded but this may only be visible on a large scale in the field.

How Formed

Gneiss is formed by intense metamorphism caused by the high temperatures and pressures deep (several kilometres) within the earth. Both sedimentary and igneous rocks may be subjected to this type of metamorphism, giving rise to quite a variety in mineralogy.

Mineral Content

Often the mineral content of gneiss is much like that of granite: mostly quartz and feldspar and often mica and pyroxene or amphibole. Accessory minerals may include garnet and kyanite.

Test: Gneiss breaks across or oblique to banding since there is no schistocity (tendency to break with a wavy surface of mica minerals) to the rock.

Interesting Facts

People talk about granite in the Precambrian Shield of Canada, whereas in actual fact many of the rocks of the Shield are granite gneisses. Within the Shield the Grenville and Churchill geological provinces were metamorphosed or "reworked" several times, giving rise to granite gneisses, whereas the Superior geological province was quite stable, thus it includes many granitic bodies.

Uses: Decorative building stone is a present day use of many types of gneiss.

The Acasta Gneiss is the oldest know crustal rock. It has been dated as four billion years old. It is a quartz-feldspar gneiss coming from near the Acasta River, east of Great Slave Lake and 350 km north of Yellowknife.

Similar Rocks

Granite (p. 248) has a similar mineral content to many types of gneiss but lacks the banded texture of gneiss. In granite the crystals or grains have a random orientation while gneiss has definite segregation of minerals in bands.

Schist (p. 286) has more continuous bands than gneiss, which allows it to be split along planes of mica whereas gneiss may be broken across the layers.

Amphibole-mica gneiss (R649): Passa San Giacomo, Ossola Valley, Piedmont, Italy. Width of specimen: 15 cm Photo: R.A. Gault

MARBLE

Marble is one of the best known metamorphic rocks because it is widely used in building materials and art. Since this rock is mainly composed of calcite or dolomite, try the vinegar test to determine the presence or absence of calcite and/or dolomite.

Appearance

Colour: Usually marble is pure white but it may be streaked or patchy with grey, yellow, green, or red.

Texture: A medium- to coarse-grained rock. Often the grain size is not evident due to uniform colour. Original fossils or sedimentary textures are usually destroyed during metamorphism.

How Formed

Marble results when heat and pressure are applied to limestone or dolostone. The fine grains of the limestone are completely recrystallized into coarser, interlocking crystals. The wisps and swirls of other colours in the marble are due to impurities of sand, clay, or iron oxides.

Mineral Content

Marble contains mostly calcite (p. 138), sometimes considerable amounts of dolomite (dolomitic marble), and rarely serpentine, graphite, and diopside.

Test: The acid test on some powdered material is the best. A strong fizz indicates calcite and weaker fizz dolomite. Marble is soft and easy to cut. It often weathers to dark grey, almost black.

Interesting Facts

The word marble is derived for the Greek, *marmaros*, meaning shining stone, because marble is quite translucent. Light can enter to a depth of several millimetres, then it is scattered back out of the rock giving it an apparent luminosity or waxy appearance. This feature has made marble the choice of sculptors for the past 2,000 years.

Uses: Greeks were among the first people to use marble as a building material. The Temple of Artemis in Ephesus, Turkey, dating to 350 BCE, has 125 enormous, 20 m marble columns. The Greeks continued their architectural supremacy with such marble structures as the Parthenon, the Theseum, and the Temple of Zeus. The ancient Greeks were also the first to bring marble into the home bathroom where it is still used today.

There are many more mundane, but indisputably serviceable, uses of marble. As a source of calcium carbonate, it is used in cement, plastics, paint, and toothpaste.

Similar Rocks

Quartzite (p. 296) is much harder than marble, resisting scratching with a knife while marble is easily scratched.

Limestone (p. 276) can in some cases resemble marble but it is usually much finer-grained. Limestone formations such as travertine and stalactites (p. 278), have a distinct layering that marble does not have.

Marble (R528): Marble showing coarse-grained calcite cleavages. Newcomb, New York. Length of specimen: 20 cm Photo: R.A. Gault

QUARTZITE

Quartzite is a tough, durable rock resistant to erosion, thus often forming ridges and hilltops. These geographical features are usually void of vegetation because little soil is available since quartzite does not erode readily.

Appearance

Colour: Usually white to grey but sometimes contains iron impurities that make it red or pink.

Texture: Individual grains are not easily visible in quartzite. It is difficult to break and tends to splinter. Breaks will be across the grain boundaries. No sedimentary features are visible.

How Formed

Quartzite results from thermal and pressure metamorphism of sandstone. The metamorphism is intense enough to weld or fuse sand grains and matrix together into a coherent, hard rock.

Mineral Content

Quartzite contains mostly quartz (p. 218) and sometimes minor amounts of feldspar, mica, ilmenite, and garnet.

Test: The rock breaks with difficulty and tends to send shards flying when hammered. It is very hard.

Interesting Facts

The 2.5-billion-year-old quartzite deposit that extends along the north shore of Lake Huron, Ontario, is known as the La Cloche Mountains. The name is derived from a legend that the Aboriginal people used these rocks to send signals. When struck, the quartzite produces a ringing sound that travels a great distance like a bell, hence the French word "la cloche." These hills or "mountains" have been immortalized in paintings by members of the Group of Seven: Franklin Carmichael and A.Y. Jackson.

Uses: Quartzite is hard, so it would make a good building material. But because of its dull appearance, granite is usually chosen instead. Its hardness and durability makes quartzite a good material for railroad bed, on which to lay track.

Similar Rocks

Marble (p. 294) is much softer and easier to crush.

Sandstone (p. 272) breaks between sand grains in the matrix.

Quartzite with jasper (R639): Welded, durable quartzite.
Lawson Quarry, Sudbury District, Ontario.
Length of specimen: 21 cm
Photo: R.A. Gault

SERPENTINITE

Serpentinite is a general term for rocks that are mainly composed of serpentine minerals (p. 214). Most commonly, it forms as an alteration product of iron and magnesium silicate minerals. Serpentinite indicates very specific types of geological environments.

Appearance

Colour: Usually apple to dark green or greyish green to almost black. Often the colouring is mottled with patches of dark and light green. The weathered surface is typically brownish to beige.

Texture: Serpentinite may be fine-grained, fibrous, or platy but most often there are no grains visible. The rock may feel slippery and appear glassy or waxy.

How Formed

Serpentinite results from alteration of rocks such as peridotite, basalt (p. 262) or gabbro (p. 252). The alteration takes place in the presence of hot hydrothermal solutions that change

Serpentinite (R941): Freshly broken surface of the same specimen figured on p. 299. Note the cream-coloured, weathered crust.

Serpentinite (R941): Weathered serpentinite Asbestos.
Shipton Tp., Richmond Co., Québec.
Width of field of view: 21 cm

olivine and pyroxene to serpentine. These conditions are met at the collision line of two continental plates; thus it may be inferred that serpentinites are an alteration product of either mantle rock or oceanic crust rock.

Mineral Content

Serpentinite contains mostly serpentine minerals (p. 214) and sometimes minor amounts of magnetite, chromite, chlorite, and talc.

Test: It may feel slippery due to inclusion of talc and chlorite. The identification of fibrous serpentine minerals would be conclusive.

Interesting Facts

Serpentine geological belts are evident in the landscape as they do not support plant growth. Soils are thin or absent, they are low in plant nutrients such as calcium and potassium and they are rich in elements toxic to plant growth such as nickel, chromium, and magnesium.

Serpentine belts are common in the world but it is curious how few of them contain economic deposits of asbestos.

Uses: Serpentinite is the main source of serpentine minerals which have many industrial uses (p. 215). As a rock it is used as a decorative building stone. It is well known as the carving stone of Inuit people.

Similar Rocks

Basalt (p. 262) is harder than serpentinite because of the glass content.

Gabbro (p. 252) is coarse-grained while serpentinite does not have grains evident.

GLOSSARY

Adamantine – a non-metallic, brilliant, shiny and diamond-like lustre.

Alloy – a mixture of two or more metals

Amorphous – lacking an atomic structure

Batholith – a large body of intrusive rock that crystallizes at considerable depth within the earth.

Bladed – the descriptive habit for flattened crystals.

Botryoidal – having the rounded forms of a bunch of grapes

Cleavage – the tendency of crystals of minerals to break in certain definite planes so as to yield more or less smooth surfaces. Cleavage is determined by the arrangement of the atoms in the mineral as it is the plane of weakest atomic bonding.

Chatoyant – describes the lustre seen in the changing luminosity of a cat's eye. It is usually caused by inclusions of fine, parallel fibres; i.e., tiger's eye quartz, cat's eye beryl or star ruby.

Conchoidal – a rounded, shell-like surface formed on a mineral when fractured by a blow.

Concretion – curved growth layers having a common centre.

Contact metamorphism – the alteration of rocks adjacent to a heat source like a magma chamber.

Crystal – a regular, repeating arrangement of atoms, which often displays symmetric, planar faces.

Desert rose – a rosette formation of gypsum or baryte crystals that have grown in and included sand.

Ductile – capable of being drawn out into wire or thread, like gold.

Element – one of a class of substances that cannot be separated into simpler substances by chemical means. They are organized into the periodic table of the elements.

Extrusion or extrusive – an igneous rock derived from a magma that has poured out or ejected at the earth's surface.

Felsic – a rock composed of light-coloured minerals such a feldspar or feldspathoids (*fel*), quartz (silica hence *sic*) and muscovite mica. Opposed to **mafic**.

Geode – a hollow concretionary or nodular stone often lined with crystals

Hackly – (of fracture) rough or jagged, as if hacked

Hydrothermal – pertaining to warm to hot, aqueous solutions occurring within or on the surface of the earth.

Igneous rock – a rock solidified from a silicate melt.

Intrusion or intrusive – an igneous rock that is emplaced in an older rock.

Kimberlite – a rock composed largely of olivine and phlogopite mica forming at depths with sufficient pressure to produce diamonds.

Lamellar (of cleavage) – breaking into thin, flat plates.

Massive – (of mineral specimens) – having no outward crystal form, although crystalline in internal structure.

Malleable – capable of being extended or shaped by hammering or by pressure from rollers.

Mafic – a rock composed of dark-coloured minerals rich in magnesium (*ma*) and iron or ferrous (*f*) minerals such as pyroxene, amphibole and olivine. Opposed to **felsic**.

Metamorphic rock – a rock derived from the alteration of a pre-existing rock by changes in temperature, pressure or chemical composition.

Mineral – a chemical element or compound, with a crystal structure and formed through a geological process.

Opaque – Impenetrable to light; not transparent or translucent

Ore – a metal-bearing mineral or rock, or a native (*i.e.* naturally-occurring) metal, which occurs in sufficient quantity to be valuable enough to be mined.

Pinacoid – a crystal form defined by two parallel faces.

Placer – a surface gravel and sand deposit containing particles of gold or other valuable mineral. The deposit forms from the mechanical action of rivers or winds or glaciers.

Polymorph – a mineral having the same chemical composition as another mineral, but with a different atomic (crystal) structure.

Porphyry – an igneous rock that has large grains or crystals within a much fine-grained matrix.

Precipitation (mineral crystallization) – the process by which a mineral crystallizes out of solution. This could be in magma or in water.

Prismatic crystal – a crystal form having three or more faces whose intersections are parallel lines.

Pseudomorph – *pseudo* 'false', *morph* 'form', a crystal that has grown with a particular crystal form may be replaced by a different mineral species, which keeps the form and dimensions of its predecessor.

Pumice – a volcanic rock, usually rhyolite, that is glassy and full of holes left by gas bubbles.

Regional metamorphism – a large scale metamorphic event involving both heat and pressure. Example: the rock alteration at the collision of two continental plates.

Resinous – resembling resin, the sticky substance exuded by some coniferous trees.

Rhombic – that of a rhomb- or diamond-shaped crystal face.

Rhombohedral – a crystal form having six identical diamond-shaped faces. These faces or planes are symmetrical to each other. The carbonate minerals calcite, rhodochrosite and siderite commonly have this form.

Rock – an aggregate of mineral grains (usually more than one mineral species) that are cemented or fused together.

Schistocity – a planar or folded texture consisting of lamellar or elongate minerals.

Sectile – capable of being cut smoothly with a knife.

Sedimentary rock – particles of rock or mineral that originate from weathering or erosion and are subsequently transported, deposited and cemented or a rock composed of mineral particles precipitated from a solution or from secretions of organisms.

Skarn – a metamorphic rock that is usually variably coloured green or red, occasionally grey, black, brown or white. It usually forms by chemical alteration of rocks during metamorphism and in the contact zone between magmatic intrusions like granites with carbonate-rich rocks such as limestone or dolostone.

Skeletal (hopper) – during rapid crystal growth the edges of the crystal may develop perfectly but leaving the core only partially filled. Examples are bismuth, gold, halite, galena, quartz.

Tabular crystal – a crystal form that is described as flattened with respect to two parallel faces.

Translucent – light is able to pass through the crystal but no clear image is transmitted; like frosted glass.

Vitreous – used to describe a lustre that resembles reflection coming from glass.

Vug – a cavity in a rock that is often filled or partially filled with a mineral that is either massive or crystalline.

TABLE OF EARTH'S CHEMICAL ELEMENTS

Element	Symbol	Atomic no.	Discoverer	Year	Other Name
Actinium	Ac	89	Debierne/Giesel	1899/1902	
Aluminum	Al	13	Wöhler	1827	
Antimony	Sb	51		—	stibium
Argon	Ar	18	Rayleigh and Ramsay	1894	
Arsenic	As	33	Albertus Magnus	1250	
Astatine	At	85	Corson et al.	1940	
Barium	Ba	56	Davy	1808	
Beryllium	Be	4	Vauquelin	1798	
Bismuth	Bi	83	Geoffroy the Younger	1753	
Boron	B	5	Gay-Lussac and Thénard; Davy	1808	
Bromine	Br	35	Balard	1826	
Cadmium	Cd	48	Stromeyer	1817	
Calcium	Ca	20	Davy	1808	
Carbon	C	6		—	
Cerium	Ce	58	Berzelius and Hisinger; Klaproth	1803	
Cesium	Cs	55	Bunsen and Kirchoff	1860	
Chlorine	Cl	17	Scheele	1774	
Chromium	Cr	24	Vauquelin	1797	
Cobalt	Co	27	Brandt	c.1735	
Copper	Cu	29		—	cuprum
Curium	Cm	96	Seaborg et al.	1944	
Dubnium	Db	105	Ghiorso et al.	1970	
Dysprosium	Dy	66	de Boisbaudran	1886	
Einsteinium	Es	99	Ghiorso et al.	1952	
Erbium	Er	68	Mosander	1843	
Europium	Eu	63	Demarcay	1901	
Fermium	Fm	100	Ghiorso et al.	1953	
Fluorine	F	9	Moissan	1886	
Francium	Fr	87	Perey	1939	
Gadolinium	Gd	64	de Marignac	1880	
Gallium	Ga	31	de Boisbaudran	1875	
Germanium	Ge	32	Winkler	1886	
Gold	Au	79		—	aurum
Hafnium	Hf	72	Coster and von Hevesy	1923	
Helium	He	2	Janssen	1868	

Element	Symbol	Atomic no.	Discoverer	Year	Other Name
Holmium	Ho	67	Delafontaine and Soret	1878	
Hydrogen	H	1	Cavendish	1766	
Indium	In	49	Reich and Richter	1863	
Iodine	I	53	Courtois	1811	
Iridium	Ir	77	Tennant	1804	
Iron	Fe	26		—	ferrum
Krypton	Kr	36	Ramsay and Travers	1898	
Lanthanum	La	57	Mosander	1839	
Lead	Pb	82		—	plumbum
Lithium	Li	3	Arfvedson	1817	
Lutetium	Lu	71	Urbain/ von Welsbach	1907	
Magnesium	Mg	12	Black	1755	
Manganese	Mn	25	Gahn, Scheele, and Bergman	1774	
Mercury	Hg	80		—	hydrargyrum
Molybdenum	Mo	42	Scheele	1778	
Neodymium	Nd	60	von Welsbach	1885	
Neon	Ne	10	Ramsay and Travers	1898	
Nickel	Ni	28	Cronstedt	1751	
Niobium	Nb	41	Hatchett	1801	columbium
Nitrogen	N	7	Rutherford	1772	
Osmium	Os	76	Tennant	1803	
Oxygen	O	8	Priestley/Scheele	1774	
Palladium	Pd	46	Wollaston	1803	
Phosphorous	P	15	Brand	1669	
Platinum	Pt	78	Ulloa/Wood	1735/1741	
Plutonium	Pu	94	Seaborg et al.	1940	
Polonium	Po	84	Curie	1898	
Potassium	K	19	Davy	1807	kalium
Praseodymium	Pr	59	von Welsbach	1885	
Promethium	Pm	61	Marinsky et al.	1945	
Protactinium	Pa	91	Hahn and Meitner	1917	
Radium	Ra	88	Pierre and Marie Curie	1898	
Radon	Rn	86	Dorn	1900	
Rhenium	Re	75	Noddack, Berg, and Tacke	1925	
Rhodium	Rh	45	Wollaston	1803	

Element	Symbol	Atomic no.	Discoverer	Year	Other Name
Rubidium	Rb	37	Bunsen and Kirchoff	1861	
Ruthenium	Ru	44	Klaus	1844	
Samarium	Sm	62	Boisbaudran	1879	
Scandium	Sc	21	Nilson	1878	
Selenium	Se	34	Berzelius	1817	
Silicon	Si	14	Berzelius	1824	
Silver	Ag	47	Prehistoric	—	argentum
Sodium	Na	11	Davy	1807	natrium
Strontium	Sr	38	Davy	1808	
Sulfur	S	16		—	
Tantalum	Ta	73	Ekeberg	1801	
Technetium	Tc	43	Perrier and Segré	1937	
Tellurium	Te	52	von Reichenstein	1782	
Terbium	Tb	65	Mosander	1843	
Thallium	Tl	81	Crookes	1861	
Thorium	Th	90	Berzelius	1828	
Thulium	Tm	69	Cleve	1879	
Tin	Sn	50		—	stannum
Titanium	Ti	22	Gregor	1791	
Tungsten	W	74	J. and F. d'Elhuyar	1783	wolfram
Uranium	U	92	Peligot	1841	
Vanadium	V	23	del Rio	1801	
Xenon	Xe	54	Ramsay and Travers	1898	
Ytterbium	Yb	70	Marignac	1878	
Yttrium	Y	39	Gadolin	1794	
Zinc	Zn	30		—	
Zirconium	Zr	40	Klaproth	1789	

Index

Page numbers in *italics* refer to figures or captions
Page numbers in **bold** refer to main entries

T